UNRAVELING   DNA

UNRAVELING DNA

The Most Important Molecule of Life

OOOOO

Maxim D. Frank-Kamenetskii

Revised and Updated

Translated by
Lev Liapin

ADDISON-WESLEY
Reading, Massachusetts

Quote from Erwin Schrödinger on page 27, by permission of Cambridge University Press.

Quote from Warren Weaver on page 122, by permission of Rockefeller Archive Center, Sleepy Hollow, New York.

Library of Congress Cataloging-in-Publication Data

Frank-Kamenetskii, M. D.
   [Samaia glavnaia molekula. English]
   Unraveling DNA : the most important molecule of life / Maxim D. Frank-Kamenetskii : translated by Lev Liapin. — Rev. and updated.
     p. cm.
   Includes index.
   ISBN 0-201-15584-2
   1. DNA.   I. Title.
QP624.F6913 1997
572.8'6—dc21

                                97-15391
                                CIP

Addison-Wesley is an imprint of Addison Wesley Longman, Inc.

Jacket design by Suzanne Heiser
Text design by Dede Cummings
Set in 11-point Dante by GAC/Shepard Poorman

1 2 3 4 5 6 7 8 9 10-MA-0100999897
First printing, September 1997

Find us on the World Wide Web at
http://www.aw.com/gb/

CONTENTS

Contents

Contents

PREFACE

O F ALL THE PHENOMENA AROUND US, THE MOST PUZZLING IS
life itself. We have become accustomed to its unfailing ubiquity and
seem to have lost the ability to marvel. But go to a forest and look at
the trees, the flowers, the grass, the birds, and ants as if you are seeing them for
the first time, and you will feel awe in the presence of life's great mystery. Is
there really some common element that unites all living things, whether man
or microbe? What is it that predetermines life's continuity and eternal renewal
from generation to generation? Only those of us living in the late twentieth
century have been lucky enough to find out the answers to these age-old
questions. Actually, the answers turned out to be fairly uncomplicated, and
truly fascinating. Their essence and derivations are the subject of this book.

Central to the new science of molecular biology—which is called upon
to answer the eternal question of "What is life?"—is the deoxyribonucleic
acid (DNA) molecule. It is also the protagonist of this narrative. We discuss
DNA from different viewpoints, placing a special emphasis on the physical
and mathematical aspects, which makes this book different from many others written on DNA and molecular biology.

This book has a history of its own. It first appeared almost fifteen years
ago in Russian, under the title *The Most Important Molecule*, as part of the
Quantum Library science-popular series published by Nauka. The book was
enthusiastically hailed by both the general public and the scientific community in the (now former) Soviet Union. Over 150,000 copies of the Russian-
language edition sold out in a short time. The book turned out to be especially popular with high school and college students interested in modern
biology, physics, and chemistry. The second Russian edition, substantially

revised and updated, appeared in 1988 and sold 130,000 copies; translations of this edition were also published in Slovak, Georgian, and Italian. In 1993, the first English edition (again extensively revised and updated), was published by VCH Publishers in New York. The French edition (translated from the 1993 English edition) was published by Flammarion in 1996. This is the second English edition, and during its preparation, the book has once again been extensively revised and updated.

Wherever possible, I have tried to avoid using scientific terms, although it is quite impossible to dispense with them entirely, since the very basis of life is formed by a large number of complex molecules, and you cannot describe them without naming them. A glossary has been provided at the back of this book in order to help the reader with terminology. The book's chapters are largely independent of one another, which permits random reading. Those impatient to get to biological questions related to DNA may skip chapters 3, 8, 9, and 11.

Since the publication of the first English edition, several breakthroughs of paramount significance have occurred in the field. As a result, AIDS is no longer a major threat, and very promising routes to fighting cancer are in sight. These new developments are covered in chapters 10 and 12.

This book could not have been written without the constant help and support that my late wife Alla (1940–1985) gave me in preparing the first Russian edition, which came to constitute the core of the book. A special tribute is due to the inestimable contribution of Vera Chernikova, who has been my dedicated Russian-language editor and taught me the secrets of writing popular science. Larissa Panyushkina, an editor from Nauka Publishers in Moscow, was extremely helpful in preparing the two Russian editions. The first English-language edition took, with interruptions, nine years to prepare. My translator, Lev Liapin, put his whole soul into the English text and displayed infinite patience during the entire process. Charles Doering, Edmund Immergut, and Christine Irizarry, editors from VCH Publishers in New York, were of great assistance in publishing the first English edition. Lastly, I am grateful to Lisa Adams from the Boston Literary Group and Heather Mimnaugh, who was an editor from this publishing house, for arranging the current edition.

—Maxim D. Frank-Kamenetskii
March 1997, Boston Massachusetts

CHAPTER 1

From Modern Physics to Modern Biology

ooooo

*Very remarkable things are happening in biology. I think that Jim
Watson has made a discovery which may rival Rutherford in 1911.*
—FROM THE LETTER OF MAX DELBRÜCK TO NIELS BOHR,

DATED APRIL 14, 1953

The 1930s

During the first thirty-odd years of the twentieth century, the most spectacular revolutionary changes were happening in physics. The advent of the
relativity theory and quantum mechanics shook the old science to its very
foundations, giving it a fresh, powerful impetus for further development,
both in depth, to explore the universal laws of nature, and in breadth, to
make further advances in related fields.

Ernest Rutherford's discovery of the atomic nucleus in 1911 proved to
be one of the major milestones in the emergence of a new physics. The very
stability of Rutherford's atom spelled the demise of classical physics. It was to
be replaced by modern quantum physics, which explained the stability of
atoms and their striking line spectra.

Originated by Max Planck, Albert Einstein, and Niels Bohr, the quantum
theory was given a remarkably clear formulation in the famous Schrödinger

equation in 1926. Beside resolving all the puzzles of atomic spectra, quantum mechanics laid the groundwork for theoretical chemistry. At long last, the riddle of the atomic number in the Periodic Table of the Elements was pierced. Quantum mechanics explained why atoms stuck together to form molecules (i.e., it disclosed the nature of chemical bonding and valency).

By the early 1930s, physicists felt they were omnipotent. The atoms had been dealt with, as had the molecules. What else was there left to do? Oh yes, there was still the atomic nucleus to look into, and so they got down to that. "Well, the nuclear problem is quite a challenge," the pacesetters reasoned, "but hardly enough to keep everybody busy. We have to come up with something clearly more substantial." Their wandering gaze lingered and stopped on the mystery of life, the Holy Grail previously shunned by physicists. Could modern physics solve that mystery? And what if the phenomenon of life was found to be incompatible with quantum mechanics? This would mean that other laws of nature would have to be discovered. What a tempting prospect!

About that time, Max Delbrück, an up-and-coming German theoretical physicist, was looking for a job he would enjoy. He had tried quantum chemistry, then nuclear physics—both quite interesting pursuits, but not fascinating enough for him. Then, in August 1932 while visiting Copenhagen, he heard Bohr lecture at the International Congress on Light Therapy. In his address, "Light and Life," Bohr shared his thoughts on the phenomenon of life taking into account the latest breakthrough in physics. Despite his complete ignorance of biology at the time, Delbrück was so impressed by Bohr's talk that he instantly and firmly resolved to devote himself to biology. Back in Berlin, he had the good fortune to meet Nikolai Timofeeff-Ressovsky, a Russian geneticist.

Delbrück began to invite his physicist friends to his home, including Timofeeff-Ressovsky among them. The Russian would speak for hours on end, explaining his science—genetics. In the process, he would pace to and fro across the room like a lion in a cage. He would discourse on Gregor Mendel's mathematically precise laws governing heredity, on genes, and on Thomas Morgan's remarkable works that proved that genes were arranged in a chain in chromosomes—tiny worm-shaped bodies inside cell nuclei. He spoke of the small fruit fly, *Drosophila* and the mutations (i.e., genetic alterations) induced in it with X rays, having studied these mutations in collaboration with the physicist Zimmer.

Delbrück was highly intrigued by the work of these men and the field of genetics. In fact, he found that the most striking thing about genetics was its breathtaking similarity to quantum mechanics, which had introduced into physics the concept of discreteness—the notion of jumps—and also had forced recognition of the role played by chance. Biologists, however, had also discovered a discrete, indivisible particle—the gene—that could randomly jump from the "ground state" (geneticists term it the *wild type*) to "excited" or "mutant states."

What is a gene? What is it made of? These were the subjects frequently debated at Delbrück's home seminars. Timofeeff-Ressovsky would argue that, in general, this question was of little interest to geneticists and that they regarded the gene in the same way that physicists considered the electron: as an elementary particle.

"Let me ask you," he once said when pressed for an answer about the gene's structure, "what is the electron made of?" Everybody laughed. "Now you see, geneticists do the same when asked what the gene is made of. The question of what a gene is is beyond the scope of genetics, and it is pointless to address this question to geneticists," Timofeeff-Ressovsky continued. "And it is for physicists to find the answer."

"Well, but still," Delbrück pressed on, "are there some hypotheses, even if purely speculative?" Pausing to think for a moment, the Russian exclaimed: "Of course there are! My teacher Nikolai Konstantinovich Koltsov, for one, believes that a gene is a polymer molecule, one of protein most likely." "And what does this explain?" the tall and lean Delbrück would prod the brawny, vigorous Timofeeff-Ressovsky further. "Will calling genes proteins help us to understand how they reduplicate? That is the main problem, isn't it? You yourself told us how the peculiar lip shape repeated itself from generation to generation in the Hapsburg family for many centuries! What was it that was responsible for such an exact reduplication of genes? What was the mechanism behind this, that operated unfailingly for many centuries? Does chemistry offer us similar examples? I, for one, have never heard of anything of this kind. No, we need a completely new approach, for in fact, this is a real mystery. A great mystery. Possibly, a new law of nature. The chief problem now is how to approach it experimentally."

Thanks to Timofeeff-Ressovsky, Delbrück was beginning to understand something of genetics. At least he no longer felt baffled by the devilish terminology that appeared to have been devised to frighten off the uninitiated. Previously, listening to geneticists, he would wonder why they needed to invent such gibberish. Weren't they rogues? His original thought was that it is criminals who use their special jargon to conceal their dark designs from innocent people. After becoming acquainted with Timofeeff-Ressovsky, however, he had changed his mind. The notorious phrase "a recessive allele affects the phenotype only if the genotype is homozygotic," with which geneticists liked to baffle laymen, now struck him not only because of its perfect clarity but because of its newly discovered elegance. "I bet it could not have been put any better," he now thought.

The Phage Group

Delbrück fell captive to the great mystery locked up in the short word *gene*. How do genes double, or reduplicate, when a cell divides? Delbrück was especially excited to learn of the existence of bacterial viruses, or the so-called bacteriophages (literally, "bacterium eaters"). These mysterious particles, which could hardly be described as living organisms, behaved just like large molecules when taken outside the cell—they could even form crystals. But twenty minutes after such a particle has penetrated a bacterium, the cell envelope splits and 100 exact copies of the original virus spill out. It occurred to Delbrück that bacteriophages were much more suitable for the study of the process of gene reduplication than bacteria, let alone animals. By studying bacteriophages, one could eventually understand the gene's structure. "Here lies the clue," Delbrück thought. "This is a very simple phenomenon, much simpler than the division of a whole cell. It should not be too difficult to understand. In fact, one only has to see how external factors will affect reproduction of virus particles. One also has to conduct experiments in different temperatures, different media, and with different viruses."

In this way, a theoretical physicist metamorphosed into a biologist–experimentalist; however, his mentality and his ideology had remained that of a physicist. What counted most was the goal. Delbrück thus appeared to be the only person in the world who was studying viruses with the sole objective of understanding the physical nature of the gene.

Delbrück left Nazi Germany in 1937, a year that was a milestone in many respects. In that year, the Rockefeller Foundation began to fund projects involving the application of ideas and methods of physics and chemistry to biological sciences. Warren Weaver, the foundation's manager, came to Berlin to invite Delbrück to the United States to devote himself entirely to the study of the problem of bacteriophage reduplication. Weaver, a physicist by education, was clearly aware of the importance of the work Delbrück was doing. (Incidentally, Weaver was the first to call the new science *molecular biology*.) Naturally, Delbrück quickly accepted the offer, since life in Germany at that time had become unbearable.

In America, Delbrück gathered a handful of enthusiasts who had been taken with his idea of studying heredity using bacteriophages. This gave rise to the Phage Group. Years went by, and the group amassed extensive knowledge on the process of phage infection, the impact of external factors on the reproduction of phage "offspring," and so on. Many remarkable studies were conducted, especially in the area of the mutation process in bacteria and bacteriophages. It was Delbrück's work during this period that earned him his Nobel Prize many years later. However, all these studies did not seem to push the researchers any closer to resolving the main question regarding the physical nature of the gene.

As frequently happens in science, people who get together to grapple with a major task gradually begin to apply themselves to the painstaking study of issues of secondary importance, and, having established themselves as authorities in their selected narrow fields of research, lose sight of the beckoning summits. Thus, travelers moving nearer to the splendid mountains they have beheld from afar enter the forest-grown foothills that obstruct the view of the summits—forests that offer many lures, like berries, mushrooms, and other distractions. Gradually, if they wander in the foothills long enough, the alluring snowy summits seen from the distance begin to seem like a mirage. Yes, those were most likely only clouds that looked like snowy mountains. And even if they were mountains, what's the rush? It is so nice being down here in the almost virgin, untrodden forests.

Such was the case with the members of the Phage Group. They had wandered from their original goal.To remind the wayfarers of their primary purpose, the admonitory voice of a leader was needed. That voice came from Erwin Schrödinger, author of the master equation of quantum mechanics.

Erwin Schrödinger

Mountains of popular science and other literature have been written on the history of the creation of quantum mechanics. Central to all these publications (and rightfully so) is the giant figure of Niels Bohr. But take any textbook on quantum mechanics and you will see that Schrödinger's equation is the alpha and omega of this science. Like any other science, quantum mechanics has evolved through the efforts of many remarkable scientists. It is clear that Erwin Schrödinger was influenced by Louis de Broglie's epoch-making idea about waves of matter; but the decisive step was still made by Schrödinger. He was the one who synthesized all that had been done before him, performing a feat of remarkable intellectual courage and force.

Although Schrödinger's name was not as well known to the broad public as those of Einstein and Bohr, he enjoyed the profound respect of physicists and chemists. In 1944, Schrödinger published a small book under the intriguing title *What Is Life?* that described the link between modern physics and genetics. At first it failed to generate any particular interest. The war was still raging, and most of those to whom the book addressed itself were wrestling with scientific and technical problems that were vital for defeating Hitler's Germany. When the war ended, however, many scientists, especially physicists, had to start again from scratch. They were looking for a niche in the new, peaceful, scientific pursuits. It was these people who found Schrödinger's book most handy.

Above all, the book gave a very clear and concise description of the foundations of genetics. It afforded physicists a unique introduction, through the brilliant exposition of their illustrious colleague, to the substance of a science that, despite its murky and baffling terminology, still held a mysterious attraction. Moreover, Schrödinger popularized and developed the ideas of Delbrück and Timofeeff-Ressovsky on the link between genetics and quantum mechanics. As long as these ideas were put forward and defended by people unknown to physicists, they tended to be dismissed as unworthy of attention. But now that Schrödinger himself was talking about them, it made all the difference.

All those who attacked the gene problem in subsequent years, including the protagonists of the unfolding drama—James Watson, Francis Crick, and Maurice Wilkins—admitted that they were greatly stimulated by Schrödinger's book. Schrödinger thus turned out to be the man who gave the rallying cry: "There they are, those glowing summits! They are right before you! What more are you waiting for?"

X-Ray Crystallography

Of the many places where Schrödinger's voice was heeded, two were destined to play crucial roles—the celebrated Cavendish Laboratory in Cambridge, England, once headed by Rutherford himself, and King's College in London. Here the final scenes of the drama were played out, the denouement being the discovery of the physical nature of the gene.

The locale of both places was not accidental. At that time (the early 1950s), Britain's scientific school of X-ray crystallography was the world's best, and this technique gave physicists a unique tool for probing the mysteries of life.

Quantum mechanics provided the theoretical foundation for understanding the internal structure and properties of matter, from atoms and molecules to a piece of iron and the crystal of ordinary salt. However, the variety of structures that atoms can form is virtually boundless. Theory alone was of little assistance in identifying the structure of a particular material. One could, of course, conjecture, but nothing could be asserted with confidence, for the conceivable variants were just too many. Therefore, a need existed for an experimental technique that would directly identify the atomic structure of matter. X-ray crystallography was precisely such a technique.

Everybody is familiar with X rays. If you are suspected of having a broken leg or of having contracted pneumonia, your physician will most likely have X rays taken to be sure. The X ray's physical nature is the same as that of visible light or radio waves. They are all variants of electromagnetic radiation, and differ only in wavelength. The wavelength of X rays can be as short as 10^{-10} m. The distances between atoms in molecules and crystals are on the same order of magnitude. These facts led the German physicist Max von Laue to assume that X rays passing through a crystal with a regular arrangement of atoms will create a diffraction pattern much like the one created by visible light passing through a diffraction lattice.

Experiments first conducted in 1912 fully confirmed Laue's guess: a beam of X rays passing through a crystal created a peculiar yet regular array of spots on film placed behind the crystal (Figure 1). Soon it was discovered

7

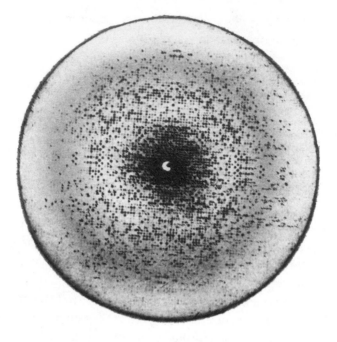

Figure 1. An X-ray pattern of a protein crystal.

that the diffraction pattern revealed precise information about the arrangement of atoms in the crystal and, in the case of crystals made up of molecules (molecular crystals), even provided information enabling the inner structure of the molecules to be determined. X-ray analysis had been born. Two Britons—Henry (the father) and Lawrence (the son) Bragg—contributed the most to the development of this technique. X-ray crystallography permitted a precise determination of the structure of all minerals and a countless variety of molecules.

Step by step, X-ray crystallographers were studying more and more complex structures. Finally, during the 1930s, some turned to biological molecules. After the first attempts, however, they balked at the enormity of the challenge. Obtaining crystals from biological molecules proved to be very tricky; when possible, the scores of thousands of atoms comprising each molecule created a diffraction pattern so complex that it defied any attempt at interpretation. It seemed impossible to reconstruct the spatial positions of these huge masses of atoms. Many years passed before such hurdles could be overcome.

These were the problems researchers at the Cavendish Laboratory wrestled with in the pre- and postwar years. The laboratory, headed by Sir Lawrence Bragg, concentrated on determining the structure of protein, which was only natural, since in those years everyone was convinced that proteins were the most important molecules of living cells. Indeed, all enzymes (i.e., molecules responsible for the various chemical reactions inside the cell) are proteins. Proteins are also the cell's main building blocks. It is not surprising, then, that everyone believed that genes too were made of protein. Therefore it was taken for granted that the road to unraveling all the mysteries of life lay in solving the puzzle of protein structure.

Protein is a polymer (a compound formed by the joining of smaller molecules), with amino acids serving as the "building bricks" or residues (Figure 2). The amino acid residues always form a strictly linear array, like a lineup of soldiers. This, however, is true of both biologically active protein and protein heated up to, say, 60°C, at which point it becomes biologically inactive. The upshot is that protein's chemical structure (i.e., the amino acid sequence) is insufficient to make it biologically active. To be biologically active, the protein molecule has to have a specific spatial, three-dimensional arrangement of residues in addition to a particular amino acid sequence. Rather than being rings or spheres, which can be mistakenly deduced from Figure 2, each amino acid has its characteristic shape. The protracted struggle in the Cavendish Laboratory raged precisely over the need to determine protein's specific spatial structure on the basis of X-ray diffraction patterns of the type shown in Figure 1. It was not until the mid-1950s that John Kendrew and Max Perutz accomplished this objective by solving the first protein structure. (The structure of the gene had already been determined by that time. As it turned out, protein structure had nothing to do with that of the gene.)

Watson and Crick

Of those who responded to Schrödinger's rallying call, two proved lucky enough to beat everyone else to the summits. They were Jim Watson, a youthful fosterling of the Phage Group, and Francis Crick, an obscure, albeit not so youthful, worker at the Cavendish Laboratory.

Captivated with the desire to learn the physical nature of the gene and convinced that the Phage Group could not accomplish the task, Watson

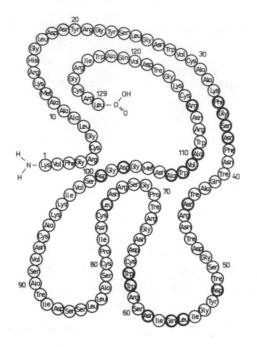

Figure 2. The amino acid sequence of a protein (lysozyme).

managed to get transferred to Europe in 1951 and wound up in the Cavendish Laboratory where he met Crick, who was equally eager to attack the gene problem. By that time, Watson had become convinced that the mystery of the gene could be pierced by determining the structure of DNA rather than that of protein.

As a matter of fact, deoxyribonucleic acid (which is the cumbersome name behind the abbreviation DNA) was nothing new. The Swiss physician Fritz Miescher had discovered it in cell nuclei as early as 1868. Since then it had been shown to be concentrated in chromosomes, which seemed to indicate its possible role as material for the gene. In the 1920s and 1930s, however, the prevailing opinion was that DNA was a regular polymer consisting of strictly repeated units of four residues (adenyl, guanyl, thymidyl, and cytidyl) and so was incapable of carrying any genetic information.

DNA was believed to play some structural part in chromosomes, and genes were thought to be made of protein, which is part of a chromosome. Therefore, what was it that prompted Watson and Crick to question the protein concept of the gene? One major factor was the work completed in 1944 by three American bacteriologists from the Rockefeller Institute, headed by sixty-year-old Osvald Avery. For many years, Avery was engaged in a painstaking study of the phenomenon of transformation, first discovered in experiments with pneumococci, the pathogens that induce pneumonia. In these remarkable experiments, two strains of pneumococci were collected, one consisting of cells capable of inducing the disease, the other lacking that ability. The pathogenic cells were killed by heating and then mixed with the living, harmless ones. It turned out that some of the living cells "learned" from the dead ones how to induce the disease. It appeared that the living cells were somehow transformed by the dead ones. This phenomenon was called *genetic transformation*. In other words, something was passed from the dead bacteria to the living. But what was it? Avery and his co-workers answered this pivotal question. Although their paper appeared in a medical journal, it attracted the attention of geneticists, physicists, and chemists, rather than physicians. The work demonstrated beyond any doubt that in genetic transformation the ability to induce disease was transferred by only one substance—DNA. Neither proteins nor any other cellular component played any part in the transformation phenomenon. This is why Avery's work is now rightfully regarded as the first conclusive evidence that DNA, not protein, is the substance of heredity, the substance of genes.

Does this amount to asserting that Avery and his colleagues had beaten Watson and Crick to the summits? Beyond question, Avery made a major step in the right direction, but he failed to get to the top. Einstein once uttered these strikingly profound words: "It is only theory that decides what we manage to observe." Avery had nothing to go by in the way of theory, and he confined himself to listing bare facts. His data, however, illuminated a glaring inconsistency in the protein concept in the gene.

Geneticists thus faced a dilemma. They had either to dismiss Avery's data or to admit that the genetic substance was actually DNA rather than protein, as they had hitherto believed. Since there was nothing in Avery's findings that could be called into question, there was no way to refute them. On the other hand, many people were not ready to give up the common concept of the protein nature of the gene. As a result, the following explanation of Avery's findings was offered: Unquestionably DNA does not contain any genes, since this is impossible; but it can trigger mutations (i.e., change genes, made, as they must be, of proteins). This explanation was an uneasy compromise, since DNA proved to be a very peculiar mutagen that persistently induced the same mutation from experiment to experiment. The known substances that induced mutations statistically had never demonstrated such "single-mindedness." This could not fail to intrigue geneticists, who had long dreamed of a mutagen capable of inducing directional mutations. Thus, for the time being, the protein theory of the gene seemed to have been saved in extremis. In the process, however, geneticists and those interested in the chemical (and physical) nature of heredity had to admit that DNA deserved serious attention.

To summarize, Avery's work had cast doubt on the common opinion of DNA as nothing more than a regular polymeric molecule assigned a purely structural part in chromosomes. It had also become clear that there was more to DNA than met the eye. But that was about all. The theory that decided what Avery had managed to observe came in the form of the structural model of DNA, proposed by Watson and Crick in 1953.

Watson and Crick had no experimental findings of their own, since no one was studying DNA in the Cavendish Laboratory. DNA studies were being conducted by Maurice Wilkins and Rosalind Franklin at King's College in London. As an object for X-ray analysis, DNA proved to be even more thankless than protein. It did not form any good crystals and yielded very poor X-ray diffraction patterns of the type shown in Figure 3. The inverse problem of X-ray analysis—that is, solving the molecule's structure, as had been attempted by Perutz and Kendrew—appeared to be impossible. Still, Wilkins and Franklin found it possible to obtain some very important parameters of the molecule. In their theoretical study of the DNA structure, Watson and Crick used these parameters, as well as the detailed data on the DNA chemical structure. As a matter of fact, their modus operandi was more like a game. They knew the structure of the separate elements, the DNA residues. Like children filling in the pieces of a jigsaw puzzle, they tried to put together a structure that conformed to the X-ray data. This "game" was to produce one of the greatest discoveries in the history of mankind.

As a matter of fact, the whole of the present narrative is devoted to that discovery. I shall gradually describe all the major features of the DNA structure and their breathtaking consequences for our understanding of the phenomenon of life. But let us first consider the substance—or rather the core—of the Watson–Crick model of DNA.

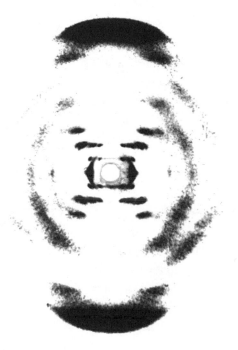

Figure 3. An X-ray pattern of DNA. Such a pattern was first obtained by Rosalind Franklin in 1953.

According to the Watson–Crick model, the DNA molecule consists of two polymer chains. Each chain comprises four types of residues—namely A (adenyl), G (guanyl), T (thymidyl), and C (cytidyl). The sequence of bases in one chain may be entirely arbitrary, but the sequences in both chains are strongly interconnected because of the complementarity principle, (Figure 4) so that

A is always opposite T,
T is always opposite A,
G is always opposite C,
C is always opposite G.

This rule of complementarity, which is the main part of the Watson–Crick model, owes its discovery largely to the data on the occurrence of different residues (i.e., nucleotides) in DNA, obtained by Erwin Chargaff.

While inside polymer chains, atoms are linked by strong covalent bonds. The two complementary strands interact through weak (the so-called intermolecular) forces. These are much like the forces that keep molecules together in molecular crystals.

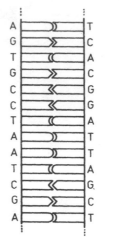

The most remarkable thing about the Watson–Crick model was the particularly elegant solution it proposed for the central problem—that of gene reduplication. In fact, to reduplicate DNA, one has only to separate its two strands and to supplement each with a complementary strand in accordance with the complementarity principle. The result will be two DNA molecules, both identical to the original (Figure 5).

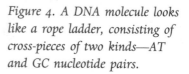

Figure 4. A DNA molecule looks like a rope ladder, consisting of cross-pieces of two kinds—AT and GC nucleotide pairs.

One can imagine how excited Delbrück was upon receiving Watson's letter containing the solution to the problem of gene structure and reduplication. It was after reading this letter that Delbrück wrote the words that serve as the epigraph for this chapter.

A lot of people besides Delbrück fell captive to the beauty of the Watson–Crick model. Geneticists who still clung fanatically to proteins as a panacea were now left with only the general argument that such a complicated thing as life could not, in its very foundation, be that simple (a flimsy argument by any standards).

Thus, DNA was recognized as the most important molecule of living nature. It was true, of course, that physicists had failed to discover any new laws of nature in biology, but their dedicated scientific quest produced the solution to the pivotal problem of biology, that of the structure of the gene.

Today, after many decades, we can say that the discovery of the DNA structure has proved as important to biology as the discovery of the atomic nucleus was to physics. Understanding the structure of the atom signaled the birth of quantum physics, and understanding the structure of DNA brought the advent of molecular biology. The parallel, however, does not end there. Pure fundamental research in atomic physics opened up the prospect of a practically inexhaustible source of energy. In a similar way, the development

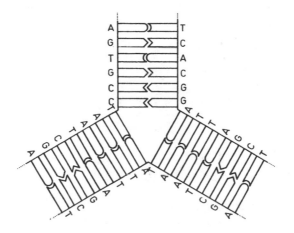

Figure 5. According to Watson and Crick, this is how the DNA replication process occurs; it results in two absolutely identical molecules being obtained from the original molecule depicted in Figure 4.

of molecular biology in recent years has opened up unprecedented possibilities for manipulating the living cell and purposefully modifying heredity. This will no doubt, for better or worse, exert an influence on humankind as radical as that of nuclear physics.

CHAPTER 2

From DNA to Protein

ooooo

How Protein is Made

Those with a pedestrian outlook often grumble that new theories pose more questions than they answer. In fact, this is so. It is not clear, however, what is so bad about this, for the more questions a new theory raises, the more valuable that theory is. Moreover, the questions raised happen to be new, have not occurred to anyone, and could not even have been formulated before the appearance of the new theory. In this respect, the Watson–Crick model of DNA may well be the absolute record holder. It is hard to imagine that in the history of science, another theory has spawned as many new questions. And some questions! They go to the very heart of the phenomenon of life. The first and foremost question was raised by the well-known theoretical physicist George Gamow in 1954.

Interestingly, Gamow's lot was much like that of Delbrück's. He was Russian and made a name for himself in 1928, when he put forward the theory of alpha decay. In 1934, after several unsuccessful attempts, Gamow fled Russia for the United States, where he lived the rest of his life. As in Germany under Hitler, life in Russia under Stalin had become unbearable by that time. His two closest friends, outstanding physicists Matvei Bronstein and Lev Landau, were both arrested during the Great Stalin Purges of 1937. Bernstein was executed, while Landau miraculously escaped execution after spending a year in NKVD, the Soviet counterpart of Gestapo, as a "German spy" (a Jew!). Landau was released due to the unprecedented and courageous interference of another great physicist, Pyotr Kapitsa, who managed to convince the murderers to "pardon" their victim. (Both Kapitsa and Landau were later awarded the Nobel Prize in Physics). If Gamow had not fled his homeland, he would certainly have become a victim of the purges.

In 1948, Gamow came forward with an idea that the universe had come into existence as a result of the Big Bang, which occurred about 15 billion years ago. He claimed that the event must have left traces in the form of radio waves still wandering through space in all directions, and predicted the spectrum of this radiation. Gamow's Big Bang theory seemed at first too extravagant. However, it was brilliantly confirmed in 1965, when two American radio engineers, Wilson and Penzias, accidentally discovered this trace radiation. Since then the Big Bang theory has gained general acceptance.

In 1954, Gamow put forward a new idea, this time in the field of biology. There are proteins, Gamow reasoned, that are the cell's chief working molecules. They are responsible for all the chemical transformations inside the cell. Almost all the building blocks of the cell are also of a protein nature. Even chromosomes consist half of DNA and half of proteins. Consequently, the work of a cell is determined by the set of proteins in it.

An individual protein molecule may contain from a few dozen to a few hundred residues. But if all the cell's proteins are separated into individual residues, only twenty kinds of amino acids will be left. The variety of amino acids as chemical compounds can be unlimited, and chemists can, in principle, synthesize any amino acid. Living nature, however, uses only twenty quite definite amino acids, which have thus been baptized as natural or canonical. This set of twenty amino acids is absolutely the same for all living things on earth. Whether you take a small bug or a big bug, you will discover that each contains the same set of amino acids. What, then, is the difference between the two? The difference lies in the sequences of amino acid residues in proteins.

And what determines the protein sequences? The answer given to this question by classical genetics was very formal: These sequences are predetermined by the genes. In what way? Classical genetics (sometimes referred to with sufficient justice as formal) could offer nothing intelligible in the way of an answer to this question.

Now, after the publication of the work of Watson and Crick, Gamow asserted, this fundamental question was answered once and for all. The amino acid sequences of all the cell's proteins are determined by the sequence of nucleotides in one of the two complementary DNA strands. These DNA residues, referred to as *nucleotides*, can be of four types (A, T, G, and C), as mentioned in the previous chapter. Information on the sequence of the twenty types of amino acid residues is thus encoded in DNA as a sequence of four types of nucleotides. Therefore, Gamow proclaimed, the cell must have a "dictionary" to translate the "four-letter" DNA text into the "twenty-letter" text of proteins! That was how the idea of the genetic code came into being.

This immediately spawned a host of questions: How does the code materialize—that is, where in the cell and by what means does the DNA text get translated into the protein language? How does DNA's long nucleotide text eventually produce relatively short protein chains? Does the DNA text perhaps consist of individual "sentences" each corresponding to one protein? Are these "sentences" the precise genes of classical genetics? What lies between the "sentences"? What fills the role of "periods" separating "sentences"? In other words, how do genes differ physically and chemically (i.e., in the molecular sense) from the space between them? Finally, what is the genetic code, this dictionary of the living cell, like?

A numerically small but determined army of scientists, scattered throughout different laboratories of the world, began to storm new summits. The invisible regiments were led by Francis Crick, the recognized leader among molecular biologists at that time. Between 1954 and 1967, all the principal questions were answered. These answers subsequently provided the basis for the central dogma of molecular biology. Not all the answers, which originally seemed to be truths established once and for all, stood the tests of the "roaring" 1970s. Although they have ceased to be dogma, these answers have until now served as the foundation for the entire edifice of molecular biology.

First and foremost, the chemical structure of DNA failed to reveal any particular characteristics that would differentiate some sections from others. Over its entire length, a DNA molecule is a continuous sequence of nucleotides of four types—A, T, G, and C. In this sense, a DNA text differs from a printed text, with its periods, commas, and intervals between words. A DNA text is an uninterrupted sequence of letters that also act as punctuation marks. These are specific sequences of nucleotides positioned between sections whose own sequence codes the amino acid sequences in proteins. An individual section of this type was called a *gene*.

A gene is thus a part of a DNA text containing information about the amino acid sequence of one protein. Now the "elementary" particle of heredity, once so hotly debated by Delbrück and Timofeeff-Ressovsky, had assumed a perfectly concrete, molecular, atomic meaning. It turned out that, rather than being an "indivisible particle," the gene was made up of hundreds of sequential nucleotides of a polymer DNA molecule, which are the elementary particles of genetic material.

How does a gene give rise to protein? This occurs in two stages. During the first, called transcription, a particular enzyme recognizes the sequence of nucleotides between the genes (this DNA segment is referred to as the promoter) and, moving along the gene, makes a copy of it in the form of an RNA molecule.

The chemical structure of a molecule of ribonucleic acid (which is what RNA stands for) is very similar to that of a molecule of DNA. It is also a polymer chain consisting of nucleotides, but unlike DNA, RNA is a single strand. Like DNA, RNA is made up of nucleotides of four types. Their chemical formulas—looking quite formidable, one must admit—are given in Figure 6. In what ways do RNA nucleotides differ from those of DNA? For C, A, and G the difference lies in the fact that in each of these nucleotides, the lower right-hand OH group changes to H in DNA (hence the prefix *deoxy*). The case of the uridine nucleotide (U) is more complex, since the passage from RNA over to DNA is accompanied not only by H replacing the OH group in the sugar, but also by the H in the top CH group being replaced by a methyl (CH_3) group in the nitrogenous base. This accounts for the difference in the names of the RNA nucleotide (uridine) and the DNA one (thymidine), although they are very much alike, both serving as partners to A in forming complementary pairs.

A gene is copied according to the same complementarity rule that governs DNA reduplication, the sole difference being that the role of T in DNA is played by U in RNA. Synthesis of RNA proceeds on one of the two complementary DNA gene strands. The enzyme in charge of the synthesis (i.e., the process of transcription) is called RNA polymerase.

Thus, RNA polymerase makes a messenger RNA (mRNA) copy of the gene's DNA. The copy is then utilized at the second stage of protein synthesis, in a process called translation. This stage is crucial, since it is here that the genetic code is brought into play.

Translation is a complex process, involving a host of protagonists, the principal one being the ribosome (a small body that functions as the cell's protein-synthesizing machinery). A ribosome is an extremely complex molecular machine, made up of about fifty different proteins and a molecule of RNA—not the RNA, that directs the synthesis of protein on ribosome, but

Pyrimidines

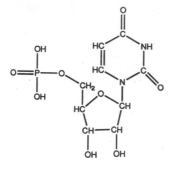

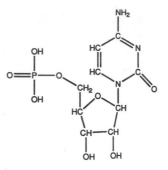

Uridine monophosphate   Cytidine monophosphate

Purines

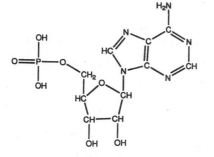

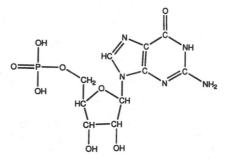

Adenosine monophosphate   Guanosine monophosphate

Figure 6. Chemical formulas of the four RNA nucleotides. At the top are the pyrimidine nucleotides, uradine and cytidine (U and C); below, are the purine nucleotides adenosine and guanosine (A and G). Nucleotides within DNA (see Figure 3.) differ in that, instead of the right-hand lower OH group (attached to the 2 carbon in the sugar), they simply have H. In addition, DNA, instead of having the uracil nitrogenous base, includes the thymine base (the top C-H group in the RNA nucleotide's ring is replaced by the C-CH₃ group).

another, ribosomal RNA (rRNA), which is an integral part of the ribosome. A ribosome acts like a molecular computer, translating texts from the nucleotide language of DNA and RNA into the amino acid language of proteins. This narrowly specialized "computer" operates according to only one program, which goes by the name of the *genetic code*.

The Genetic Code

In the late 1950s and early 1960s, Francis Crick, Sidney Brenner, and colleagues defined the main properties of the genetic code. The code turned out to be a triplet one, with each amino acid matched by a sequence of three nucleotides on RNA. The three-nucleotide sequence of RNA was called a codon. The text encapsulated in mRNA is translated consecutively, codon by codon, beginning with some starting codon, according to this scheme:

mRNA: ... AAGAAUGGAUUAUCCAACCGCCCCGUAU ...

protein: $a_0- \quad a_1- \quad a_2- \quad a_3- \quad a_4- \quad a_5- \quad a_6- \quad a_7 \quad ...$

In this scheme a_0, a_1, ... stand for amino acid residues of protein. It will be recalled that there may be twenty types of them. From this information, it can be easily calculated that there is a total of $4^3 = 64$ different codons. It can further be assumed, then, that not all the codons are matched by an amino acid. However, the unmatched nonsense codons are very few, and they specifically serve as stop signs that mark the end of a protein chain. This is why they are also referred to as *termination codons*. Additionally, there are several codons that correspond to each particular amino acid residue, which amounts to saying that the code is degenerate.

By 1961, it became clear that the genetic code is a triplet one, it is degenerate, nonoverlapping (i.e., the reading proceeds codon after codon), and contains initiation and termination codons. The next step was to identify the correspondence of each amino acid residue to specific codons and to find out which codons denote the beginning and end of the protein chain synthesis. The way to proceed was quite clear. One "only" had to read in a parallel manner two texts: the DNA (or RNA) text of the gene and the amino acid text of the protein corresponding to that particular gene. Then these two texts could be compared—and *voilà!*

It will be recalled that ancient Egyptian manuscripts were deciphered by using precisely this technique. The hitch in this case, however, was that although biologists, by the 1960s, had learned how to decipher protein sequences, a technique for reading either DNA or RNA sequences was still lacking. That was why another approach had to be taken.

Now imagine that instead of the Rosetta stone (which bears the same inscription in both Egyptian hieroglyphics and in Greek), a live ancient Egyptian had been unearthed during Napoleon's conquest of Egypt. In this case, there would have been no need for Champollion's genius to compile a French–Ancient Egyptian dictionary. One would only need to show the Egyptian different things and ask him to supply the corresponding hieroglyphics. American geneticist Marshall Nirenberg (National Institutes of Health) utilized this concept to decipher the genetic code.

The fact is that cells do know the code! Consequently, one only has to request that the cells identify different nucleotide sequences. However, one must have a clear idea of what these sequences are. By that time, researchers had learned to synthesize some (but by no means just any!) artificial RNAs. It was pointless, however, to offer such an RNA to a living cell: It would simply gobble it up (i.e., break it up into individual nucleotides and use them to build its own RNAs). That is why instead of using living cells, Nirenberg opted to use cell extracts that retained the capacity to synthesize protein but lacked the enzymes to break up RNA. The extracts were unable to do many other things a living cell could do, but the one important thing that they could do was to synthesize protein from instructions of an "alien" RNA. Such extracts were baptized as "cell-free" systems.

To an extract of intestinal bacilli, Nirenberg added an artificial RNA consisting only of uracils. The first question put to the cell-free system was what amino acid corresponded to the UUU codon: the clear-cut answer was that of phenylalanine. This finding, reported by Nirenberg to the Moscow International Biochemical Congress in 1961, caused a sensation. The road to deciphering the code had been cleared!

Similar correspondences were soon established for many amino acids. However, it proved to be difficult to identify nucleotide sequences in artificial mRNA. At that time, no one was able to synthesize even short fragments of predetermined sequences. Researchers could only obtain polynucleotides with random sequences from certain mixtures of monomers, so they began to think of other techniques for deciphering codons. The situation changed drastically, however, due to an unexpected breakthrough.

We have seen that the problem of the code had first been raised by a physicist and that the code's general properties were defined by genetic techniques, after which the biochemists took over. The problem was finally resolved when biochemists were joined by synthetic chemists. Chief among them was Har Gobind Khorana.

By 1965, Khorana had learned to synthesize short RNA fragments of a predetermined sequence—first doublets (dinucleotides) and then triplets (trinucleotides). With the help of enzymes, such doublets and triplets were synthesized into long polynucleotides in which the doublets and triplets repeated themselves many times over. After that, the polynucleotides, with a strictly definite and known sequence, were added to the cell-free system to determine their correspondence to protein chains.

By 1967, the deciphering of the genetic code (Figure 7) was finally completed. All codons that correspond to each of the twenty amino acids had been identified, as had the termination codons. But what about the initiation codons? The fact is that codons that serve an exclusively initiating function do not exist. Under certain specific conditions, this function is assumed by the AUG and GUG codons, which normally correspond to the methionine and valine amino acids.

Even a cursory look at Figure 7 is sufficient to detect a striking regularity. The degenerate nature of the code is clearly not fortuitous: that each amino acid is matched by a definite codon is determined mainly by the two first nucleotides. What the third nucleotide will be is almost immaterial. In other words, despite the code being triplet, the chief message is carried by the doublet that stands at the beginning of the codon. One can thus say that the code is quasi-doublet.

This hallmark of the code was noted in the very early stages of the deciphering exercise. You cannot, of course, encode with doublets all twenty amino acids, since the total number of different doublets can only be $4^2 = 16$. This is why we must assume that the third nucleotide in a codon does have a particular role to play.

There is, however, a rule obeyed by the code almost to the letter. To formulate it, we have to recall that, by their chemical nature, the four nucleotides belong to two different classes: the pyrimidines (U and C) and the purines (A and G). Consequently, the rule of the degeneracy of the code may

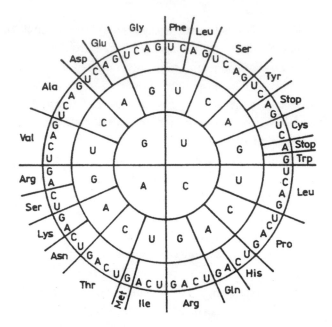

Figure 7. The genetic code. The first letter of the codon is located in the central circle; the second, in the first ring; and the third, in the second ring. Written in the outer ring are the abbreviated designations of the twenty amino acids. (Note that three amino acids—Arg, Leu, and Ser—are repeated, due to their matching with codons that begin with different nucleotides.) The Stop symbol stands for termination or stop codons.

be formulated as follows: If two codons have the same first two nucleotides and their third nucleotides belong to the same class (purine or pyrimidine), they encode one and the same amino acid.

Take one more look at the code's table and you will see that this rule is observed quite strictly, with two exceptions. If the rule were followed with unreserved strictness, the AUA codon would correspond to methionine, rather than isoleucine, and the UGA codon would match tryptophan, rather than serve as a terminating signal.

Is the Code Universal?

"Excuse me," the reader may well object, "but the cell-free system is obtained from a specific organism. Can one guarantee that code-deciphering experiments on a cell-free system taken from a different organism will yield the same result?" This is a good question. Quite naturally, it came up during the code-deciphering exercise.

Initially, researchers made sure to specify that what they were talking about was the code of *Escherichia coli*, not the code in general. The first cell-free system was obtained from this bacterium and was the object of the investigations described earlier. However, everything seemed to suggest that the genetic code of other organisms was no different from that of *E. coli*. Nirenberg repeated his experiments with cell-free systems taken from a toad and a guinea pig. The experiments failed to detect any variations from the *E. coli* code. Thus, there seemed to be no doubt that the code was universal.

It was true that some *E. coli* mutants revealed deviations from the code: Some of the nonsense codons were read as sensible (i.e., as corresponding to definite amino acids). This phenomenon was called *suppression*. However, it was clear that the genetic code structure had to be highly conservative and resistant in the process of evolution.

Let us imagine that the code undergoes a sudden, albeit a very slight, change, and one of the codons changes sense to match a different amino acid. This codon, however, is present in not one, but many genes. All of these genes will then begin to synthesize proteins with one amino acid substituting for another. Although the substitution will go unpunished for some of the proteins, which will retain their functions, one can hardly assume that no damage will occur to an important protein in every case. It is an established fact that replacing one amino acid in a single protein amounts to throwing a monkey wrench into the works, thus causing the death of the whole organism.

A classical case in point is that of sickle-cell anemia, a very severe hereditary disorder caused by the substitution of a single amino acid in a single protein hemoglobin. The hemoglobin molecule consists of four poly-amino acid chains—two identical α-chains and two identical β-chains—bonded by intermolecular forces. The sticking together of several chains frequently results in a functional protein. The mutation in question consists of glutamic acid, the sixth amino acid in the β-chain in normal hemoglobin, being replaced by valine. From the genetic code diagram, (Figure 7) one may conclude that in the codon matching the sixth amino acid of the β-chain, T in the DNA sequence was substituted for A in the second position.

This substitution modifies the structure of hemoglobin, sharply reducing its capacity to carry oxygen in the body. The name of the disorder derives from the fact that the change, occurring at the molecular level, distorts the form of oxygen-carrying cells (the red blood cells) from round to sickle shaped.

Examples such as this make it clear that the code must remain immutable in the course of evolution, which in turn means that it should be universal for all living things.

CHAPTER 3

Meet the Most Important Molecule

∞∞∞

We believe a gene—or perhaps the whole chromosome fiber—to be
an aperiodic solid.

—ERWIN SCHRÖDINGER, *WHAT IS LIFE? THE PHYSICAL*
ASPECT OF THE LIVING CELL, 1944

It Is Like . . . a Corkscrew

The blueprint for what every one of us will be like appears at the instant in which the gametes of our father and mother merge to form a single whole, called the zygote or the fertilized egg. The entire message is encapsulated in the nucleus of this single cell—more precisely, in its DNA molecule. This molecule carries information about the color of our eyes and hair, about our stature, the form of our nose, whether or not we will be a virtuoso musician, and many other things. Of course, our future depends not only on DNA but also on the unpredictable vicissitudes of life. However, many, many things in our individual destiny will be determined by the qualities built into us at birth by our genes—that is, by the sequence of nucleotides in our DNA molecules.

DNA replicates itself at each division of the cell, so that every cell encapsulates information about the entire organism. It is as if every brick of a building stored a miniature plan of the entire building. What if architects had followed such a precept throughout history! Restorers nowadays would not have to rack their brains over, say, the original appearance of the Zeus Altar of Pergamum, even if but one stone was all that remained.

The fact that a specialized cell actually knows how the whole organism is made was demonstrated spectacularly by the British biologist J. Gurdon. Extracting a cell nucleus from a frog, he used a sophisticated surgery technique to implant it into another frog's egg, from which its own nucleus had been removed. A normal tadpole or even a frog developed from the hybrid egg, and was a perfect copy of the donor of the cell nucleus. Such experiments, performed recently on sheep, created a real hullabaloo. A clone, called Dolly, of an adult sheep was grown by Scottish biologists. Nature itself sometimes creates such clones—in the case of identical, or homozygotic, twins. This happens when the two cells separate after the zygote's first division and each results in its own organism. Because the DNA molecules of each organism are entirely identical, the twins are very much alike.

What is the structure of DNA, the queen of the living cell? It is not like a simple rope ladder, as one might think after looking at Figure 4. The ladder is actually twisted into a right-handed helix, and resembles a corkscrew—a double one (although rarely does one see such corkscrews). Each of the DNA strands forms a helical line twisted right-handedly, as is the case with a corkscrew (Figure 8). The four types of nitrogenous bases, whose sequences encode genetic information, provide a sort of stuffing for the corkscrew-like cable, whose outward surface consists of two sugar–phosphate polymer chains making up DNA. The residues making up DNA are very similar to those of RNA, whose chemical structure is given in Figure 6. That is why, rather than drawing all four nucleotides once again, we shall confine ourselves to showing only what T looks like (Figure 9), which seems to differ most from U, its RNA counterpart. The upper ring is called the nitrogenous base and the five-member ring is called the sugar; the phosphate group is to the left.

What are DNA's chief parameters? The diameter of the double helix is 2 nanometers (nm), and the distance between the neighboring base pairs along the helix is 0.34 nm. The double helix turns full circle every ten pairs. The length of the DNA depends on the organism it belongs to. The DNA of simple viruses contains but a few thousand residues; that of bacteria, a few million residues; and that of higher organisms, billions of residues.

When all the DNA molecules of a human cell are straightened out into a line, the result is a string with a length of about seven feet. Consequently, the length of the line will exceed its thickness by a billion times. For a better idea of what that means, imagine that DNA is twice as thick as it is in Figure 8, that is, about 2 inches. DNA of that thickness, taken from a single human cell, would be long enough to gird the earth along the equator. Using this scale, the cell's nucleus would be the size of a stadium, and the actual human being would be the size of the earth.

Hence, packing up the DNA molecule in a cell's nucleus, especially in the case of multicellular organisms whose DNA molecules are very long, is understandably quite a problem. Moreover, it has to be packed up so as to make DNA's entire length accessible to proteins (e.g., to RNA polymerase transcribing the necessary genes).

Replication of molecules of this length poses yet another problem. In fact, following DNA's reduplication, the two complementary strands, that are originally twisted around each other, many times, have to become separated. This means that before the replication is over, the molecule will have rotated around its axis millions of times. These facts illustrate that the questions raised by the work of Watson and Crick were far from limited to the problems of the genetic code and related matters.

The questions raised also tended to generate doubts. Is the Watson–Crick model, in fact, correct? How reliable is the foundation upon which all the findings of molecular biology rest? By being highly specific and highly

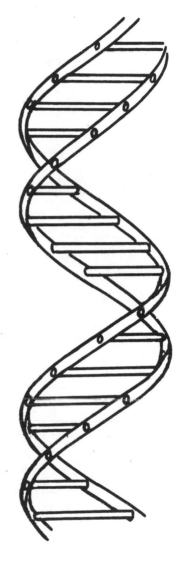

Figure 8. DNA is a ladder twisted into a right-handed helix.

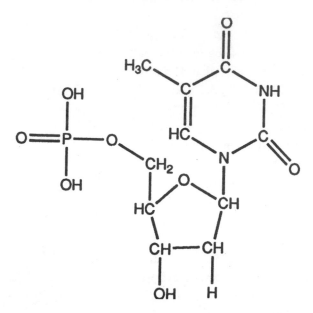

Figure 9. Thymidine monophosphate is a thymine nucleotide that is part of DNA. The remaining three DNA nucleotides have a similar structure, but each has a nitrogenous base of its own (the top group). These three bases (adenine, guanine, and cytosine) are similar in DNA and RNA (Figure 6).

detailed, the Watson–Crick model was also vulnerable. If investigators were to discover one convincing fact at variance with the model, it would be enough to topple the double helix from its pedestal. Thus, physicists set out to expose weaknesses in the model.

If every DNA molecule really consisted of two polymer chains, some reasoned, and these chains were bonded to one another by weak noncovalent forces, then an experiment could clearly register their coming apart in a heated DNA solution. If nitrogenous bases in DNA really formed hydrogen bonds with each other, others thought, this could be verified by measuring DNA spectra in the infrared range or studying the speed with which ordinary (light) hydrogen exchanges with heavy hydrogen (deuterium). If the nitrogenous bases' reactive groups were really buried inside the double helix, reasoned still others, could DNA respond to substances that react only with those groups? Experiments were conducted to answer these questions and many others. By the close of the 1950s it became clear that the model had stood the first test. Attempts at debunking it had failed, one after another.

It Is Like a Windowpane

The need to test the Watson–Crick model was not the only motive that impelled physicists to take up the study of the molecule. The molecule itself offered enough attraction.

Schrödinger's book, *What Is Life?*, contains a prophetic statement (taken as the epigraph for this chapter). DNA really looks like a solid. The base pairs in it are arranged as in a crystal. It is, however, a linear, one-dimensional crystal, with each base pair flanked by only two neighbors. The DNA crystal is aperiodic, since the sequence of base pairs is as irregular as the sequence of letters in a coherent printed text. And like letters in a printed text, the pairs of bases AT and GC have similar sizes in both width and height.

Thus, it came as no surprise that the one-dimensional DNA crystal, a crystal of an entirely new type, had very much intrigued physicists. Was it perhaps a semiconductor? Or might it even be a superconductor—and at room temperature at that? The DNA was subjected to another investigation, with the verdict being that it was not a semiconductor, and even less of a superconductor. It turned out to be a quite pedestrian insulator, like a windowpane, and to be as transparent as glass. A water solution of DNA (it dissolves very well in water) is just a transparent liquid. The analogy with glass does not end there. Ordinary glass, including a windowpane, is transparent for visible light and is a strong absorber of ultraviolet (UV) radiation. DNA is also an absorber of this region of the spectrum. Unlike glass, however, which is not harmed by UV rays, DNA can be severely damaged by them.

Penetrating DNA, a photon (a quantum of UV radiation) transmits its energy to the nitrogenous base, thus exciting it. The excited state may resolve itself variously. If the photon is absorbed by adenine, guanine, or cytosine, nothing special will happen—the absorbed energy will transmute itself rapidly into heat (as is the case with the windowpane), leaving DNA intact. The issue is quite different if the photon gets absorbed by thymine—not just any thymine, but the one neighboring another thymine in the chain. In this case, before the absorbed energy has a chance to be transformed into heat, the two neighboring thymines enter into a chemical reaction. The result is a new chemical compound called a thymine photodimer, T<>T (Figure 10).

The dimer's structure is unorthodox. In fact, carbon ordinarily is either tetrahedral, when its links with the neighboring atoms are directed from the tetrahedron's center to its vertices; or triangular, when the bonds lie in one

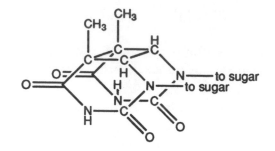

Figure 10. A thymine photodimer.

plane and are directed from the center to the vertices of the equilateral triangle. However, in a photodimer the two C-C bonds of each of the carbon atoms participating in the sticking of thymines form a right angle! Moreover, all four atoms of carbon form a quadrangle (called cyclobuthane).

Thus, damage has been inflicted on the DNA by the photon: In the place of two thymines has appeared an entirely new chemical compound that halts further progress of enzymes working on DNA. "Trained" to recognize only the letters A, T, G, and C, the enzymes balk at the mysterious T<>T newcomer they do not know. If this flaw is not expunged from the "text," the enzymes will not be able to transcribe DNA's information and synthesize RNA. All life in the cell will come to a standstill, and it will perish.

In fact, UV rays are so deadly for the DNA molecule that in the process of evolution the cell developed a special system to repair the damage done by them. The enzymes of the repair system begin the process. First, the UV-endonuclease enzyme recognizes the thymine dimer and cleaves the sugar–phosphate backbone at its location. Another enzyme then proceeds to widen the nick. This results in an enormous gap of several thousand nucleotides around the point of the formation of the thymine dimer in one of the DNA strands. The thymine dimer gets removed in the process, together with a host of normal nucleotides, just in case. This bodes no ill, since the other, complementary strand that has remained intact serves as a template for still another enzyme, the Kornberg's DNA polymerase (also known as DNA polymerase I), to rebuild the second strand and thus restore the double helix back to the normal state—identical to the DNA prior to the damage.

This is yet another important funtion of DNA's double-strandedness! It not only produces two identical copies of genetic material, but also guarantees the safety of the message encoded in DNA. If DNA consisted of one strand between the replication cycles, it would not be able to repair itself.

From unicellular organisms to man, all cells have a repair system. This is no surprise: after all, life emerged under the sun. It may appear strange that the repair system should be active even in cells that have never been exposed to solar radiation, like the cells of the bowels. A witty explanation of this has been offered by G. M. Barenboim, who proposes that DNA protects itself against the Cherenkov UV radiation arising in cells as a result of the decay of the natural abundance of radioactive isotopes.

The inactivation of the repair system as a result of mutation is a disaster. Children born with the defect called xeroderma pigmentosum show no tolerance to exposure to light; their skin develops sores, which gradually grow into malignant tumors. Despite extreme care to keep them from exposure to solar radiation, these children are doomed. (Incidentally, direct experiments on animals have demonstrated that thymine dimers can cause cancer.)

This means that lying in the sun may not be a totally innocuous pastime. We cannot, of course, deny ourselves the pleasure, but we should be careful not to overload our repair system. Besides, the repair process itself is not innocuous. The pivotal enzyme of the system, Kornberg's DNA polymerase, is thought to have a knack for erring, and thus a repair can cause mutation. For their part, somatic mutations (those occurring in the body's cells) are now believed to be a major factor causing malignant transformations of tissue (see Chapter 12).

A host of troubles springs from the simple fact that DNA is sensitive to UV radiation! And this is true with only the tiny fraction of UV rays that reaches the earth's surface, since the bulk of it is absorbed by the atmosphere. So, does one have to regret the fact that DNA is as transparent as an ordinary windowpane with respect to visible light?

It Melts, but Not Like Ice

Those who expected some unusual physical properties of DNA were eventually rewarded. The one-dimensional and aperiodic nature of the DNA crystal fully reveals itself when DNA is melted. Although DNA's crystal state appears to be crystal clear, how can one imagine its passage into a liquid state? What shape will a one-dimensional DNA crystal take when it is melted?

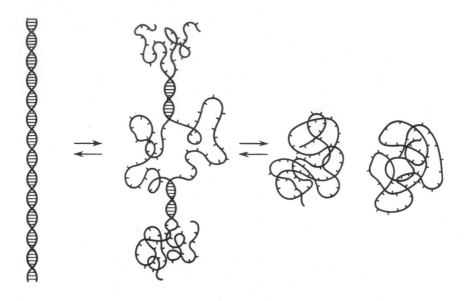

Figure 11. Melting of DNA.

Before we get down to clarifying this point, let us recall why ordinary ice melts. Ice is a crystal made up of H_2O molecules. A perfect order reigns in the arrangement of water molecules bound together by as many intermolecular bonds as possible. Some of these links are ruptured; others are distorted in liquid water. What is it that keeps water liquid in temperature 0°C? By losing some of the bonds and weakening others, water molecules gain a much greater leeway for moving and spinning, an exercise that grows increasingly "rewarding" with the temperature. As the heating up proceeds further, water molecules, to gain still greater freedom, sacrifice the last remaining bonds, thus undergoing the passage from a liquid to a gaseous state. This is a general trend: As temperature rises, substances tend to manifest a readiness to sacrifice the energy of intermolecular bonds for the sake of increased entropy.

This also holds true for DNA: A rising temperature makes the existence of the double helix "unprofitable." The intermolecular bonds that keep the two complementary strands of the molecule together rupture, and the result is two single strands (Figure 11). From the viewpoint of entropy (i.e., in the sense of gaining greater freedom), this seems profitable because, no longer attached to its complementary partner, each chain gains much greater freedom for assuming a much greater variety of configurations in space.

You cannot damage single DNA strands by simple heating: The bonds connecting nucleotides into a chain are so strong that they can be destroyed only by a strong acid or cut apart by enzymes called nucleases.

Despite a certain similarity, the melting of DNA differs drastically from that of ice. The striking difference lies in that the melting of DNA occurs at a broad range of temperatures (one of several degrees), whereas that of ice occurs at one definite point on the temperature scale—the so-called phase transition. In this transition, a substance's phase state changes abruptly with temperature—from solid to liquid, from liquid to gaseous.

We witness phase transition every day when we boil water in a kettle. In the process of boiling, the water–steam system stays on the very point of the phase transition—the kettle's temperature will never exceed 100°C as long as there is any water in it. This pattern will be repeated in heating ice or snow. The temperature will rise to 0°C, wait until all the ice has melted, and then start rising again.

Unlike the phase system, with DNA the temperature rises continuously, and as it does so, newer and newer regions of molecules pass over from a helical to a melted state. The interesting thing is that this difference is a direct consequence of the one-dimensional nature of the DNA crystal.

Physicists realized that such conduct by a substance was possible even before World War II, before DNA or any other real one-dimensional crystal was even thought of. At this time, developing a full-fledged theory of phase transitions in real three-dimensional crystals was still an intractable task (which came to fruition only much later, during the 1970s). Therefore, scientists began to wonder if it might be possible to achieve a phase transition with respect to one-dimensional or two-dimensional crystals. Analyzing the first variant proved quite simple, but there was a hitch: No phase transition could be obtained. The profound meaning of this failure was understood by the great Russian theorist Lev Landau (who we already mentioned in Chapter 2). Together with E. Lifshits, he wrote in 1938: "In any one-dimensional system there can be no phases, since they would tend to mix up with each other." For a long time this statement, known to physicists all over the world as the "Landau theorem," had been regarded as purely negative, meaning only that a one-dimensional system is a totally worthless model for a theoretical analysis of the problem of phase transitions.

35

It is likely that Landau never dreamt that some day real systems would be found to which his statement could be applied. However, DNA is very close to being such a system. The word *close* is used here because the Landau theorem applies only to strictly uniform one-dimensional crystals, whereas DNA, as we remember, is an aperiodic crystal consisting of two kinds of groups—the base pairs A·T and G·C—that differ in stability. The A·T pair is easier to disrupt than the G·C pair. That is why if one DNA has larger A·T/G·C ratio than another, it melts at a lower temperature.

Does it matter how many kinds of pairs there are—two or one as in a strictly uniform crystal? Yes, it does. This is a very interesting question that has been investigated by many researchers already directly in connection with the problem of DNA's melting. I have been engaged in extensive studies of the problem along with a number of people around the world, among them M. Azbel', D. Crothers, A. Dykhne, M. Fixman, I. Lifshits, E. Montroll, D. Poland, and A. Vedenov.

And what did all the research find? The conclusion made by Landau was upheld. In principle, this is also due to the one-dimensional nature of the system, but the reasons behind it are different from those operating in a strictly uniform crystal. The phases are absent, although not because they would tend to mix up, as Landau asserted, but because the DNA regions enriched with A·T pairs melt at a lower temperature than regions enriched with G·C pairs. For this reason, the transition into a new state under a rising temperature occurs not abruptly, but stage by stage and region by region.

A diagram showing the dependence of heat absorption on temperature for a DNA molecule's solution, instead of displaying an infinitely narrow peak characterizing the melting of ice, would reveal a multitude of peaks corresponding to the melting of a molecule's individual regions. The width of each peak must, according to the theory, correspond to about 0.5°C. An experiment fully confirmed this prediction. Figure 12 shows the stage-by-stage melting of DNA (of plasmid ColEl) containing about 6,500 base pairs.

It stands to reason that no one can measure the heat absorption of a single molecule. People usually have to deal with a sample consisting of billions upon billions of DNA molecules, but all of these have exactly the same nucleotide sequence. In all the molecules, the same temperature opens up the same regions. That is why the study of the effect on a multitude of similar molecules can give us an idea as to what happens to each individual molecule.

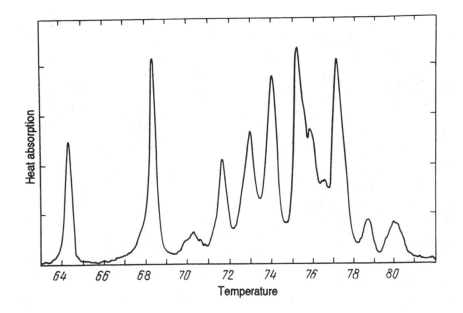

Figure 12. Dependence of DNA's heat absorption on temperature. This curve is often called the differential melting curve. The curve was obtained for DNA that has the code name ColEl and contains about 6,500 nucleotide pairs.

Members of a team from the Institute of Molecular Genetics in Moscow (Anatol Borovik and co-workers) have literally been able to see with their own eyes the stage-by-stage melting of DNA. In an experiement, they were able to register the molecule's open regions by means of a chemical agent selected specifically for that purpose. The experiment proceeded as follows. A DNA solution was heated to a certain temperature within the melting range. Individual regions of the molecule opened up in the process—the complementary strands in these regions moved apart and nitrogenous bases were exposed to a solution. Added to the solution was a substance capable of reacting with the exposed bases but not with those buried inside the double helix. Following the completion of the reaction, the sample was cooled to room temperature; the chemically modified regions were no longer able to close up again and form a double helix.

The DNA molecules thus treated were then examined under an electron microscope. One of the pictures obtained is shown in Figure 13. After taking many pictures of the molecules as they opened up at different temperatures, the researchers constructed the resulting pattern (Figure 14). Plotted on the horizontal axis is the position of a base pair along the DNA chain. The probability of the given pair being opened is plotted on the vertical axis; temperature is plotted along the third axis. A comparison with the curve showing the dependence of heat absorption on temperature (top left axis) reveals that each peak is actually matched by the melting of a definite DNA region. The figure gives an idea of the shape of the DNA molecule at every temperature

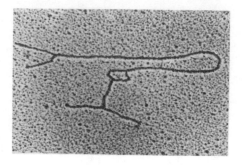

Figure 13. Photograph of ColEl DNA (using an electron microscope after its state was fixed at a temperature of 72°C). One can clearly see three diverged, melted segments—two at the ends and one in the middle.

in the melting range. One notes, for instance, that at 72°C both ends of the molecule must be melted as well as the region spaced from the left end by 60 percent of the molecule's overall length. This is quite in line with the picture in Figure 13. Note that, unlike the present case, melting in DNA does not always start with the ends. It is just that both ends of this particular molecule (ColEl DNA) are strongly enriched with AT pairs.

One can see that studying the melting of DNA has proven to be much more fascinating than studying the melting of ice. Instead of one peak whose width defies measurement, there are many peaks whose position and width are determined by the sequence of nucleotides in DNA. Each DNA has a melting profile all its own, dependent on the genetic message it carries.

The melting of DNA is much more than a unique physical phenomenon, however. It is a process that is constantly at work in the cell. Judge for yourself: In both the replication of DNA and its transcription, the complementary strands have to be separated so that both (in the case of replication) or one (in the case of transcription) can serve as templates for DNA or RNA synthesis.

How do the strands get separated? What plays the role of a "heater" that melts a DNA section? This role is assigned to specialized proteins, one being RNA polymerase. The enzyme firmly attaches itself to DNA, not in a random fashion, but to a definite sequence of nucleotides—the promoter, located between genes. Then RNA polymerase melts the promoter (opening up about ten nucleotides) and begins moving along the gene, opening on its

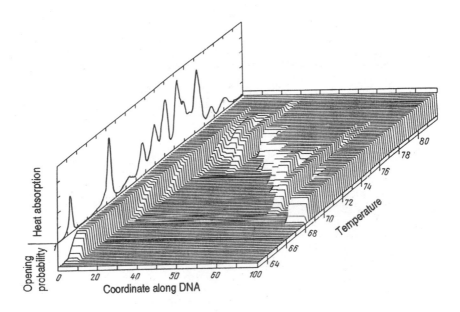

Figure 14. The full picture of ColEl DNA melting, obtained by computer processing of a large number of electron-microscope photographs of the type shown in Figure 13.

way newer regions and synthesizing the mRNA molecule. The parts of the gene that are "run over" and left by RNA polymerase shut down again while the synthesized mRNA molecule hangs down into a solution. A ribosome swims up to it to begin synthesizing protein in accordance with the laws of the genetic code. This process is shown in shorthand form in Figure 15.

The ability of DNA strands to separate and to reanneal is widely used in biotechnology and genetic engineering. Ingenious genetic engineers have invented a truly miraculous device, the polymerase chain reaction (PCR) machine, which is essentially a thermocycler that periodically heats and cools the DNA sample. In doing so, the machine actually performs the PCR, which amplifies in vitro a desired stretch of a single DNA molecule. You can literally start with a *single* DNA molecule and after n cycles of the PCR machine you will have 2^n molecules in your test tube. Thus a miracle of life, self-reproduction, is performed in a test tube.

But we have gone too far ahead in our narrative. We will speak about the advent of genetic engineering and its breathtaking achievements and

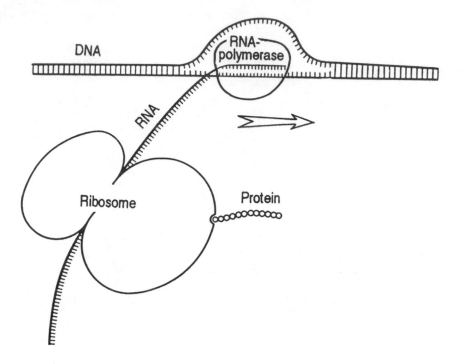

Figure 15. RNA polymerase crawls along DNA, synthesizing mRNA. The ribosome copies information from mRNA by synthesizing protein, in accordance with the genetic code.

even more breathtaking prospects in forthcoming chapters. Now let us continue to meet the most important molecule.

It Is Like the Trail of a Man Lost in the Woods

Why is a man who tries to go directly forward in the forest bound to stray in cloudy weather? Why is he doomed to perennially return from whence he came? Several superstitions exist on this score. According to some, we go round in circles because one leg is slightly shorter than the other. According to others, one step is longer than another. All this is rubbish. The actual reason is quite different. A man tries to go straight ahead, but with no reference points in the distance to guide him, sooner or later, he always deviates from the straight line. The loss of memory of the original direction occurs even faster with the increasing thickness and monotony of the forest. Rather than assuming a circular motion, the man's trail assumes an unpredictable random pattern.

To get a picture of such a trail, let us put a sheet of paper on the table and hold it down with the point of a pencil. Then, closing our eyes, let us spin the sheet around and trace a short line with the pencil. Having repeated the procedure a half dozen times, you will see on opening your eyes a broken line that, most likely, crosses itself at least once. This would give you a rough idea of a man's wandering in the forest, with the line's intersections indicating his return to where he has already been.

The man, of course, veers from his chosen direction only gradually; he does not move in zigzags unless he is drunk. The walk of a drunkard, in fact, does look like a zigzagging broken line. This is why random walks are sometimes described as drunkard's walks. (Incidentally, even if our wayfarer happens to be a teetotaler, which is not an impossible assumption, with no remote reference points to guide him, his trail in the forest will, just the same, eventually look like a random walk.)

The issue now becomes that of the length of each straight segment of the corresponding zigzagging broken line. Let us designate this length with the letter b. For a drunkard, b will be equal to one step. The next one is almost certain to be in a different direction. The teetotaler will, of course, take pains to make the b value as big as possible, but with no remote reference points, the value will still be much less than the total distance traveled if, of course, the latter is sufficiently long.

It is not only people who go astray. Molecules in a solution are also known to be "footloose": They strive to move in a straight line, but colliding with other molecules, they deviate from their paths. This phenomenon is known as Brownian motion.

The theory of random walks was proposed by Einstein. It was the subject of one of three articles, published in 1905, that determined the developmental course of physics in the twentieth century (the other two papers were on the relativity theory and light quanta). According to Einstein's theory, if a particle has covered a path L, it will have shifted from the original point to a distance $r = \sqrt{Lb}$. What does this mean?

Let us get back to the man in the thick forest on a cloudy day. The value for b (the length of a straight segment) in this instance will hardly exceed 100 feet. The speed will probably amount to about 2 miles per hour. This means that in 9 hours (when the walker will be exhausted and unable to progress further) the man will have covered a distance of a mere half a mile from his point of departure! In this period of time, not surprisingly, he will have crossed his own path many times in a vain effort to get out of the woods. The only way not to get lost is to try to increase the value of b at any price.

But what has all this to do with DNA? The link, believe me, is a most direct one. Like a man in the woods and a particle in a solution, the DNA molecule strives to straighten out into a rectilinear shape, since this position corresponds to a minimum of energy of bending. This natural urge, however, is hampered by thermal motion. Bombarded by the surrounding water molecules, the DNA molecule begins to writhe like a worm, curling itself up into an everchanging polymer coil.

That is why a double-helical DNA of sufficient length more often than not looks in solution like a coil, rather than a knitting needle. The size of the coil is described by the same Einsteinian formula $r = \sqrt{Lb}$, where L is the molecule's length and b is determined by the degree to which the DNA molecule resists thermal motion (i.e., by the bending rigidity of the double helix). Reliable measurements for DNA have shown that $b = 100$ nm.

A double helix's capacity to bend is no small biological feat. The fact is that were the DNA molecule as rigid as a knitting needle, it would never squeeze itself into a cell, not to mention the cell nucleus. Indeed, the cell, especially in higher organisms, contains many DNAs, concentrated mainly in the nucleus. If we assume that the whole of DNA in a human cell is one molecule, its length L will be about 2 m. This is a million times more than the nucleus diameter. So how does it manage to squeeze itself into the nucleus?

Might thermal motion be sufficient to pack DNA up into the nucleus? To answer this question let us estimate the diameter of a polymer coil with L = 2 m. Assuming $b = 100$ nm, it is easy to see that $r = 0.5$ mm. This will be only one thousandth of the molecule's full length, but still 1,000 times greater than the nuclear diameter. Thermal motion is thus not sufficient to squeeze DNA into so small a space.

To circumvent the difficulty, the cells of higher organisms are provided with a special mechanism for forcibly bending the double helix. Like a thread on a spool, the molecule is wound around a special set of nuclear proteins— histones. The molecule is wound twice around one "spool," then passes onto the next, and so on. A "spool" with DNA wound around it is called a *nucleosome*; DNA in the nucleus of higher organisms is like a necklace of nucleosomes. Far from forming a straight line, this necklace is neatly packed in little bodies called chromosomes. It is by this cunning ruse that the cells pull off the mind-boggling trick of squeezing a 0.5 mm-diameter polymer coil into a nucleus with a diameter of less than a micrometer.

CHAPTER 4

Under the Sign of DNA

ooooo

The Crisis of Molecular Biology

The basis of what is commonly referred to as common sense is the so-called principle of Occam's razor, which states that of all possible explanations of unknown phenomenon, the simplest one is preferred. Unwittingly or by design, this principle is followed by all sober-minded people: Granny who has lost her glasses, a detective unraveling a crime, and a scientist looking into natural phenomena. It is true that an explanation that appears to us to be the simplest may not be the correct one. And although we often realize that our choice of the simplest explanation will most likely be proven wrong, we have no alternative—the simplest explanation receives priority over the rest because it is the easiest to refute and thus needs to be checked first.

One can assert that the scientific picture of the world is a set of the simplest (for the present level of knowledge) explanations. However, the extent to which these explanations are true is a question that goes beyond the scope of today's science. In fact, what stimulates science the most is the awareness of the imperfections of our current knowledge. However, to move one step forward, such awareness alone is not enough: One also has to prove that an old notion, which seems to us so natural, is incorrect or incomplete. The unfading fascination and eternal youth of true science lie precisely in the fact that the picture of the world, as offered by science, is in a state of constant change. This is particularly true with respect to such a young science as molecular biology.

Under the Sign of DNA

In the 1950s, at the dawn of molecular biology, the question of how the DNA molecule functions in the cell seemed as easy as pie. What, after all, did one need to explain? Well, first, how does DNA reduplicate and, second, how is RNA synthesized on it? Or, to use scientific language, how do the two main processes—DNA replication and transcription—occur in the cell?

If there are two processes, they must require two enzymes: DNA and RNA polymerases. These proteins were searched for and found in the cell. So all was well. Many years later, it is true, it was discovered that the DNA polymerase that had been found (Kornberg's DNA polymerase or DNA polymerase I) was not the chief vehicle for replication. It turned out that the function of Kornberg's DNA polymerase in the cell is to repair the gaps developing in DNA during replication and repair, and that quite different enzymes are in charge of the process of DNA replication.

Luckily, no such confusion occurred with regard to RNA polymerase. It actually proved to be the enzyme in charge of transcription in the cell. However, the discovery of this enzyme fell short of providing answers to all the questions involved in RNA synthesis. The fact is that each time an RNA copy is made, it is made not from the whole of DNA but from a small section of it, containing one or several genes. So what happens to the other genes? If they lie low, then why? Could it be that there are many, rather than one, RNA polymerases whose duty is to "read" their respective genes? Perhaps other proteins exist (let us term them *repressors*) that prevent an RNA polymerase from reaching the silent genes and reading them. Which of the two explanations shall we settle for?

Let us not rack our brains in vain. In the study of living nature, it is very common for two or even more explanations to coexist and to be applicable in different cases. This is what happened with the problem of the regulation of transcription—both assumptions proved equally valid.

A repressor protein was found in *E. coli*, which attaches itself very firmly to DNA at the start of a particular gene, between the promoter and the initiating codon, thus preventing the RNA polymerase from transcribing that gene. This was how one possible explanation (the latter, above), proposed by French scientists F. Jacob and J. Monod, materialized.

Then came the second explanation. When an *E. coli* is infected with a bacteriophage, some of the genes of a phage DNA are at first transcribed by the "host" RNA polymerase. Then a totally different phage RNA polymerase comes on the scene and begins to read the rest, the so-called late genes of the

45

phage DNA. Thus, "power" changes hands in the infected cell—wrenched from the rightful ruler, the E. coli DNA, by the invader—the phage DNA. The fact that the RNA synthesis switches over from the early to the late genes was first discovered by the Russian scientist Roman Khesin and his colleagues in the late 1950s. The transcription of DNA into RNA and the closely related problem of protein synthesis directed on ribosomes by RNA were the central problems of molecular biology in the 1950s and 60s. At that time, the process of replication was thought to be as clear as day. What else was there to figure out in regard to DNA?

And so, at the close of the 1960s, molecular biologists began to say that, since DNA had been dealt with—as was, admittedly, also the case with protein synthesis (since the genetic code had been deciphered by that time)—it was time to switch to new problems, such as that of the higher nervous activity of the brain. Some, incidentally, did precisely that. Later, with the benefit of hindsight, the period could be reassessed as one in which the old ideas and methods had outlived their raison d'être, while new ones had yet to appear. To many, however, it seemed that there were no problems to look into. The most facile answers had been elevated to the rank of absolute truths. All this, incidentally, had become clear only much later (oh, the insight of hindsight!), whereas during the period in question no one could suspect that the 1970s and 1980s, and even the 1990s, would pass under the sign of DNA.

It will be recalled that the basic principles of molecular biology, then regarded as unassailable, could be briefly summarized as follows. All living organisms on earth have an identical structure of the key cellular apparatus of protein synthesis. The structure of the apparatus is this: The genetic message is stored in the form of a nucleotide sequence in a linear DNA molecule. DNA may be divided into continuous sections (genes), each encoding the amino acid sequence of one protein. Genes are separated by control sections that attract RNA polymerase and repressor proteins. They cannot overlap or be interrupted by other sequences. An RNA copy is transcribed from the gene's "start" site, which is then used for synthesizing protein on ribosomes in accordance with the universal genetic code. Thus, there is a strictly unidirectional flow of information in the cell: DNA → RNA → protein.

Individual principles of this scheme, or rather the central dogma of molecular biology, were proven on different objects, but the famous E. coli was, of course, the chief "testing site." However, the scheme was so simple and natural and seemed to interpret all the genetic data so well, that its universal applicability to all living things raised not the slightest doubt in anyone's mind. Some differences, naturally, were expected to crop up in the study of higher organisms. Thus, it was assumed that, with the higher organisms, the greater part of DNA would be constituted by the "managerial apparatus" (i.e., the control sections accompanying the genes would be much longer than those in bacteria).

"All is well that ends well," the reasonable reader would judiciously conclude. "The scientists have done a good job, and, through concerted efforts, have clarified all the main issues—from the structure of the DNA molecule down to how it operates in the cell. Now it is time for them to join hands once again in order to come to grips with the applied problems. For it is just impossible for such progress in understanding the profound processes of life to fail to bring about stunning successes in their desired modification."

Truer words were never spoken. The trouble was that, although everything seemed clear in principle, no practical application of the knowledge amassed was yet in sight. There was, naturally, a wealth of speculations about gene transplantation and genetic engineering. Things, however, did not go any further than that. Attempts to undertake something practical were blocked by the lack of the necessary techniques for cutting DNA into pieces, reshuffling those pieces, and stitching them together to suit the experimenter's design. Without these techniques, any talk of genetic engineering would remain pure Manilovism (meaning "smug complacency, inactivity, and futile daydreaming"; from Manilov, a character in Gogol's *Dead Souls*).

It is actually very easy to break a DNA molecule into pieces. As a matter of fact, it is much more difficult to avoid this, especially if it is exceptionally long. An accidental breakage, however, is not what an experimentalist is after. It was necessary to learn to induce breaks in the uniform molecules' selected sites with pinpoint accuracy, not exceeding the size of one nucleotide. But for that to occur, one would need to have a surgical knife capable of cutting a molecule with an accuracy of up to a billionth of a meter! This would be like trying to cut a sausage with such uncanny accuracy as to give a slice of it to every inhabitant of the world. Thus, accurately breaking a DNA molecule seemed to be a hopeless task.

Physicists and chemists began looking through their arsenals in an attempt to come up with an answer to this problem. "What if we hit it with a laser?" was one suggestion. Another suggestion: "Could we melt it, but only slightly, and then apply an enzyme that splits only DNA single strands? For the fact is that all the molecules with the same sequence should melt in the same places." This idea seemd plausible, and so they got down to business. It turned out that DNA could be cut in this way, but different molecules tended to vary, though slightly, in length. The variation amounted to several dozen nucleotides, and so the methodology devised was unable to solve this problem and thus fell short of meeting the stringent demands of genetic engineering.

Thus, the advent of the golden age of genetic engineering seemed to be postponed indefinitely. And it was not simply a matter of genetic engineering. The problem of cutting DNA into pieces blocked the solution of another problem—that of determining the nucleotide sequence. Despite the glib talk about promoters and other regulatory sections, genes, and the like, none of the DNA pieces, in fact, had been sequenced. The genetic code, thus, still amounted to little more than an attractive painting that could be a treat to the eye and hung on a wall in the lab. The code was the dictionary for translating texts from the nucleotide language of DNA into the amino acid language of protein. But it was precisely the DNA texts that were unavailable!

Relatively short polymers, such as proteins, had been sequenced, but DNA still stubbornly refused to divulge its sequence, first and foremost because of its length. If it could be broken up into short sections of 100 or 200 nucleotides each, it would somehow be possible to read the sequences in them. But how can you break down a long chain into small bits in a definite manner? The same vicious circle! Again one had to reach the same uncanny accuracy of up to one nucleotide! We can, thus, see that molecular biology was trapped in a blind alley.

At the close of the 1960s and the dawn of the 1970s, only a handful of cranks in the world thought it was too early to wind up fundamental research on DNA; that the current notions, despite their apparent logic and completeness, were only pale shadows of the true solutions; and that nature was infinitely more complex and fascinating. Nobody listened to them or took them seriously. Those turning a deaf ear objected: "What else, in heavens' name, do you wish to see added to this already practically perfect edifice? Of course, some minor details have yet to be clarified, but they will add nothing new in principle. Well, then, if you wish to persist in wasting your time, you may go on occupying yourselves with trifles. We, however, shall try to find something more important to occupy our time with. As regards the cutting of DNA, this, of course, is a worthy but obviously impossible task. There's no use beating your heads against the wall."

The Breakthrough

Against this backdrop, an event occurred that quickly changed the prevailing mood. The event, which marked the beginning of a new era in molecular biology, was the discovery in 1970 of reverse transcriptase—an enzyme that synthesized DNA from RNA (i.e., something like a reversed transcription). Since previously everything seemed to fit in quite nicely without the enzyme, the idea of it even existing tended to be dismissed. Now it turned out, however, that this enzyme was very real.

Was the reverse flow of information (i.e., from RNA to DNA) possible in a cell? The discovery raised quite a hullabaloo, triggering talk of the collapse of all the foundations of molecular biology, of the possibility of synthesizing RNA on the basis of protein, of inheritance of the newly acquired characteristics, and God knows what else. Furthermore, since reverse transcriptase had been discovered in viruses capable of inducing cancer in animals, it seemed obvious that from the discovery of reverse transcription it was only a stone's throw to solving the problem of cancer.

Years passed, however; the hullabaloo died down; and reverse transcriptase occupied its own rather modest place among the other enzymes. No, there is no reverse flow of information within the cell. It is just that with some viruses (including HIV, the inducer of AIDS) it is RNA, not DNA, that is the genetic substance. These viruses are supplied with reverse transcriptase to permit, upon penetration of the cell, synthesis of viral DNA within the cell.

The area in which reverse transcriptase did prove indispensable was in genetic engineering. It is with the help of this enzyme that scientists were able to obtain DNA on RNA isolated from human cells, in order to insert that DNA into a bacterial or yeast cell and stimulate it to produce, for example, interferon or other proteins needed in medicine. But we shall elaborate on that later.

Although the discovery of reverse transcriptase was important, its psychological impact was of much greater importance, since the discovery demonstrated that the principles of molecular biology were not nearly as inviolable as most people had thought. Fresh sensations soon began to appear. In the course of the 1970s, whole new classes of enzymes that worked on DNA, whose existence had not even been suspected, were discovered. These enzymes gave biologists unheard-of possibilities for interfering in genetic processes, by providing the basis for new techniques whose absence had retarded the development of molecular biology toward the close of the 1960s. Now a giant step forward had been taken amid the collapse of the seemingly unshakable notions of the structure of genes of both viruses and higher organisms (with bacteria as the only survivors). Genetic engineering had come into being—an applied branch of molecular biology.

The enzymes that made the largest contribution to the new revolution in genetics were called *restriction endonucleases*. As with reverse transcriptase, there was no place for these newly discovered enzymes in the logically completed edifice of molecular biology that had been built by the end of the 1960s. It was true that somewhere in the backyard of the edifice there loomed the vague question of the role of DNA methylation. This modification consisted of the addition of a methyl group (CH_3) to the nucleotide. But, then, nobody said that all the finishing touches had been put in place and the rubble removed; it was just that there was a lack of volunteers to do the painstaking and thankless job of clarifying the remaining immaterial details. And even if such volunteers were found, who would be willing to fund a project that promised only unrelieved boredom? The fact is, to be able to engage in a scientific investigation you are supposed to indicate in advance what you plan to discover and when. It is said that the late Howard Temin, who discovered reverse transcriptase at the University of Wisconsin–Madison and eventually received the Nobel Prize (together with David Baltimore from MIT) for his discovery, had a lot of troubles before the completion of his long quest for the enzyme. He was barely able to obtain a "grace period." What if his work had been delayed?

Let us assume that someone has undertaken to clarify the role of methylation in DNA in five, or even three, years. In order to get the necessary financial support, such an investigation, which clearly holds no promise of major fundamental discoveries, must at least spawn some eventual spin-off into, say, agriculture or medicine. But this is already downright ridiculous, for what practical applications can be expected from an investigation into DNA methylation of bacteriophages?

Happily, however, the curiosity of scientists is inextinguishable. And the problem of methylation, although seemingly quite secondary in importance, still provided sufficient food for thought. It was discovered that some nucleotides in DNA were modified chemically after the completion of replication.

The intriguing thing is the very small number of methylated groups in DNA—1 per 1,000. This means that the enzyme in charge of the process—methylase—must be able to recognize some specific nucleotide sequences. Another interesting fact is that if methylase has been "put out of action" (through mutation), the phages maturing in the bacterium turn out to be noninfectious. Such a phage will attach itself to the bacterial wall, but the DNA that it injects into the bacterium ends up being "dissolved" in the cell.

What is the reason for this? It was found that the bacterium uses methylase to "earmark" DNA of the bacteriophages that have matured in it—just like a shepherd marks his sheep. Unlike the shepherd, however, the bacterium does this to its own detriment. In fact, an "earmarked" phage is far from being a humble sheep. Once inside the cell, it turns into a deadly killer. It is not quite clear what makes a bacterium engage in marking. With no mark, however, a phage is in for a tough time. Like a shepherd who would hardly permit a sheep with a wrong mark or none at all to stay in his herd, the bacterium immediately destroys the "unbidden guest." The bacterium uses as its "weapon of destruction" certain enzymes that recognize the same sequences as do the methylases. If the sequences are found to be unmethylated, the cell enzymes proceed to tear the DNA molecule to pieces, rendering it biologically inactive.

It was the quest for an answer as to how a bacterium destroys an intruder virus that produced the discovery of the restriction endonucleases. A restriction endonuclease is an instrument that nature itself seems to have intended for genetic engineering. Since different bacteria have different ways of marking their own DNA, there are enzymes that recognize the most diverse nucleotide sequences. The isolation of these enzymes makes it possible to cut DNA into the desired pieces and then splice them in a manner to suit the experimenter's design. The result are chimerical (also know as recombinant or hybrid) molecules consisting of DNA fragments isolated from different organisms. These bits are then spliced with DNA ligase, a nonspecific enzyme capable of curing breaks in a DNA chain. The many exciting things made possible by restriction enzymes will be the subject of subsequent chapters.

CHAPTER 5

We Can Reshuffle the Genes!

ooooo

Humanity's Age-Old Dream

The period between the 10th and the 5th millennia B.C., when domestic animals and crops were first raised, was perhaps the most important in humanity's history, and determined the further development of civilization. It was domestic animals and crops that spared people the daily quest for food and induced the transition to a sedentary way of life, with all the ensuing social, economic, and cultural implications.

Little information has reached us on how the selection work operated throughout the many centuries. We may assume that selection expertise was being improved and handed down from generation to generation. We know only that even today, in our hectic age, the work of a person who develops new breeds requires enormous patience and pertinacity. As a rule, it is only after decades of painstaking everyday work that he or she reaps a harvest, usually in the declining years; many die before their efforts begin to bear fruit!

By the time human beings began to interfere with living nature, it had already traversed a long path of evolution, whereby the branches of the tree of life had long parted in different directions and developed in seemingly total independence from one another. Nature had seen to it that these different branches (species) did not get intertwined: Crossbreeding between different species either is impossible or fails to beget offspring that can reproduce. Thus, you cannot crossbreed a cat and a dog; and a mule—a crossbreed between a donkey and a horse—although it is a living creature, is incapable of reproducing itself.

This "taboo" imposes severe limitations on selection work. Essentially, plant- and stock-breeders are forced to reshuffle the same genes with only slight variations. Imagine yourself going into a shop to buy a deck of cards and discovering that they sell only decks in which all cards are the same (one deck consisting of only the seven of spades, another of only the queen of clubs, and so on), and all the differences inside each deck boil down to some cards being printed more clearly than others, some being slightly spotted, and so on. To top it off, each deck has an individual back pattern, so you cannot mix the cards of different decks. This is analagous to the position of breeders who essentially have to reshuffle the same genes. So one can only admire the remarkable results that they have been able to achieve within such tight constraints.

How much freer the creative work of a breeder would be if it were not for the barriers between the species! Countless amateur breeders have made many attempts to overcome these barriers and produce remarkable new hybrids. One such hybrid, existing only in the fiery imagination of enthusiasts, is a plant with potato tubers and tomato tops. Such alluring projects were once very much in vogue in Russia under Stalin. It was even reported that a hybrid of cabbage and radish had been made. Everything about it was fine, including a set of chromosomes and the capacity for reproduction. The only hitch was that the plant had the roots of a cabbage and the top of a radish. For many years this case provided material for satirists and humorists.

Thus, gene reshuffling has been the same "bee in the bonnet" to man as his desire to transform some substances into others (the alchemists' philosophical stone). It is no coincidence that fairy tales and myths abound of people turned into animals and back again and of places densely populated with mixed-species creatures like centaurs, fauns, Pegasuses, mermaids, and sirens.

The true magic of today's science lies in its growing ability to make age-old fairy tales and myths a reality. Nuclear physics made it possible to transform some chemical elements into others. Molecular biology has overcome the taboo on crossbreeding between species. The alchemists' dreams of gold now strike us as outdated and naive when compared to the present's virtually unlimited possibilities for producing energy and, ironically, the terrifying potential for destroying all life on earth, which stem from our acquired ability to convert some elements into others in nuclear reactors and bombs. Similarly, centaurs and mermaids seem to us to be minor compared with what genetic engineering—a new technology developing right before our eyes—holds in store for humanity. It makes possible the reshuffling of genes of organisms infinitely removed from each other in the course of evolution (e. g., a human being and a bacterium).

Genetic engineering logically grew out of the whole DNA science. However, it was the discovery of the restriction endonucleases that proved to be the breakthrough permitting the direct reshuffling of the genes. Restriction endonucleases recognize the short nucleotide sequences and proceed to cut the DNA molecule in that place. Such sequences may be encountered in any DNA. That is why, for instance, if we handle DNA of, say, a fly and DNA of an elephant simultaneously with the same restriction endonuclease, we may well trigger an accidental reshuffling of the genes of the fly and the elephant. To obtain chimerical molecules, one only has to add the DNA ligase that splices DNA fragments.

However, it is one thing to create a chimerical DNA molecule in vitro and another to make it biologically active and capable of multiplying in the living cell, while modifying its genetic properties in the bargain. Herein lies the chief problem of genetic engineering. We hasten to stress that the problem is still far from its final resolution. Furthermore, the progress in genetic engineering has been plagued by totally new difficulties that were not even suspected when the work was started. Outside of these many difficulties, however, nature had also prepared for genetic engineers a remarkable present in the form of very specific organisms called plasmids. Most of the accomplishments scored by genetic engineering to date have been connected with plasmids.

Plasmids

When Joshua Lederberg discovered plasmids in the early 1950s, nothing seemed to indicate that they were destined for a great future. Generally speaking, all that Lederberg had hit upon was that, in addition to the main DNA, which normally did not change cells, E. coli happened to contain other small DNA molecules (which he called plasmids), that the bacterial cells seemed only too willing to exchange. The discovery of plasmids in bacteria originally did not generate any special interest, because higher organisms, in addition to their main, nuclear DNA, have smaller DNA in cytoplasm.

Ironically, it was physicians, rather than molecular biologists, who were the first to signal the significance of plasmids. In 1959, Japanese doctors found that the ineffectiveness of antibiotics as a cure for dysentery with some patients was due to the fact that the bacteria with which the patients were infected carried a plasmid containing several genes of resistance to different antibiotics. It was discovered that genes of resistance to antibiotics (i.e., those that have thwarted many efforts to control bacterial infections in recent years) are always carried by plasmids. An ability to move freely between bacteria enables the plasmids carrying such genes to spread rapidly among bacteria immediately in the wake of a broad application of an antibiotic. The staphylococcus infection, now a veritable scourge of surgical clinics, owes its diabolic resistance to the plasmids.

Such notoriety has made plasmids the object of close scrutiny by both physicians and molecular biologists. A careful study of plasmids produced the conclusion that these are independent organisms of a totally special type. Previously, it was believed that the simplest objects of living nature were viruses. Viruses always consist of nucleic acid (commonly DNA, sometimes RNA) encapsulated in a protein coat. Outside the cell, a virus is just an association of complex molecules. An isolated virus particle is thus more like an inanimate object than a living creature (which, incidentally, was clearly demonstrated well before World War II when crystals were grown out of viruses). However, having penetrated into a cell, a virus comes "alive," becoming a crafty and very dangerous predator. It begins to interfere actively with the cell's functioning. Redirecting the cell's resources to meet its own needs, the virus eventually destroys the cell and multiplies a hundredfold in the process. What, one would think, could be more perfect and simpler at the same time?

Outside a cell, a plasmid is simply a DNA molecule. Inside, it leads quite an "intelligent" existence, drawing on a portion of the cell's resources for its multiplication, while taking care to restrain its voracity to keep the cell alive. In this respect, a plasmid's behavior is wiser than that of a virus, for by causing the death of the cell, the virus is doing a disservice to itself. A plasmid, on the other hand, multiplies together with the host cell. If a virus may be likened to an insatiable predator, a plasmid is more like a domestic animal, especially a pet dog.

Just as a person may have one or several dogs, and sometimes none at all, a bacterium may also have one or more plasmids, or none at all. In a favorable external environment, all these cells are about equal, but having plasmids is a bit more taxing. Like dogs, they have to be fed. If conditions suddenly change, placing the cell in a hostile environment—say, one with penicillin—the plasmid, like the faithful dog, throws itself on the enemy. Penicillinase, the enzyme produced by the plasmid, ensures the cell's survival by destroying the penicillin. That is why the coexistence of a plasmid and a bacterial cell is an alliance based on mutual advantage—symbiosis, to use a biologist's term.

The master can give one of his dogs to a friend. Similarly, bacteria can exchange plasmids. Plasmids' propensity for "promiscuity," which is quite a nuisance to physicians, has proved to be a boon for genetic engineers. Extracting plasmids from bacteria, inserting alien DNA in them, and then mixing the resulting hybrid plasmids with bacterial cells affords at least part of the hybrids an opportunity to multiply successfully in bacteria. In other words, thanks to the extreme simplicity of its structure, plasmids are actually quite receptive to alien genes being inserted into them. More complex organisms, even viruses, react much more painfully to such surgical interference.

Restriction endonucleases are used to obtain hybrid plasmids containing fragments of DNA from any organism. The hybrid plasmids are then multiplied together with the host bacterium, which makes it possible to multiply the inserted section of an alien DNA many times over. This procedure is called *cloning*. With the help of plasmids, molecular biologists can clone any DNA section. The technique has provided them with a unique opportunity for manipulating not only the genes of bacteria and viruses but also of higher organisms. This paved the way for fresh, remarkable discoveries that will be the subject of coming chapters. However, the chief goal of genetic engineering is to learn to obtain in the cells of one species the end products of genes of a different species (i.e., proteins).

Bacteria Produce the Protein We Need

One can graft into a plasmid a DNA section taken from any donor—say, a human gene—and this plasmid, once inside a bacterium, will begin producing

56

protein that corresponds to the human gene. It is a trick genetic engineers have learned to perform with the dexterity of a juggler, using one of the several techniques now in existence.

The first technique was popular in the mid-1970s during the dawn of genetic engineering, when plasmids mainly served as the recipients for the genes of *E. coli* or other bacteria. The technique is quite simple. The DNA whose genes are to be built into a plasmid, is broken into fragments at random (it is not necessary to use restriction endonuclease in this first step). Then the randomly fragmented DNA is mixed with a plasmid cut by a restriction endonuclease in one place, and DNA ligase is added. Different plasmid molecules capture different DNA fragments, resulting in a mass of different plasmids. This mixture is introduced into bacterial cells.

The chief problem with this approach is the selection of the right strain carrying a plasmid with the desired gene. Given proper criterion for such a selection, the technique can yield good results. However, although this method gave experimentalists some valuable strains that produced particular bacterial proteins, it still earned itself the name of the "shotgun" technique. In fact, it reminds one of shooting a shotgun with one's eyes closed. This early technique of genetic engineering assigned an overly important role to chance—random fragmentation, random insertion. Attempts to use the technique to obtain strains producing proteins of higher organisms failed dismally.

More recently, two specific techniques have been adopted that produce results that gain much media attention. The first involves extraction from a cell of a mRNA that corresponds to a particular protein. Through the mediation of a reverse transcriptase, a DNA copy, called cDNA, (i.e., the needed gene), is transcribed from the mRNA. Chemical techniques are then utilized to stitch to the gene the required control sections, and the resulting gene is inserted into a definite site of a plasmid. The plasmid used in the process is one designed specifically for genetic engineering. Such a plasmid has everything necessary to ensure its existence in a bacterial cell, including a promoter section from which an RNA polymerase can begin transcribing the desired gene, which is inserted directly after the promoter.

There is yet another technique—that of the direct chemical synthesis of DNA's nucleotide sequence derived from the known protein source. Given the code's degeneracy, there may be many different sequences of gene allowing the experimentalist a freedom of choice. The synthesized gene, with control sections stitched to it, is inserted into the plasmid.

The plasmid carrying the artificial gene is then incorporated into bacterial cells. Selection of bacteria with the necessary plasmids proceeds as follows. The desired gene is inserted into the plasmid together with a gene of resistance to one antibiotic; sometimes the desired gene may be inserted with a whole selection of genes of resistance to several antibiotics at the same time. The cells are grown in a medium containing such antibiotics. This technique ensures the selection of required bacteria and enables them to keep the artificial plasmids. Techniques also exist that enable each cell to store not one or two but thousands of plasmid copies. These techniques permit fantastic productivity in the output of protein encoded in the incorporated gene. Cases have been reported in which the mass of the protein amounted to almost half the cell's protein.

The development of a technology enabling the bacterial cell to produce any protein in large quantities marked a new stage in the scientific and technological revolution—the era of biotechnology. Above all else, however, this new technology revolutionized molecular–biological research itself. Usually, a specific protein is produced by a cell in very small quantities, sometimes a mere one or two molecules per cell. As a result, the production of proteins needed for particular research was an arduous and costly undertaking. One had to process dozens of kilograms, nay tons, of biomass to obtain milligrams of protein. Despite such meager quantities, it was still not possible to ensure the necessary purity of the protein. Hence, the cost of many protein preparations was exorbitant and their purity was substandard.

Genetic engineering brought about a radical change in this situation. Genetic-engineering strains now exist—superproducers of many proteins with high standards of purity—that were undreamed of before. Molecular biology supply firms have sharply diversified the production of enzymes and other protein preparations and have reduced the prices of these products. Thus, molecular biology received a powerful new impetus, resulting in an unheard of acceleration in the pace of scientific research.

CHAPTER 6

DNA Texts

ooooo

More about the Crisis

A paradoxical situation had developed in molecular biology in the late 1960s. Techniques for identifying amino acid sequences in proteins (the first protein, insulin, had been sequenced by Frederick Sanger in the early 1950s) had, by then, become rather well developed. Newer texts were being added to the protein sequences data bank. The genetic code, the dictionary for translating DNA texts into the language of proteins, had been deciphered. The irony of the situation, however, was that biologists had failed to read a single DNA text!

One could, of course, try to read parts of a text by the so-called reverse motion, relying on the protein sequences as points of departure. But this type of reconstruction was made ambiguous by the code degeneracy. More importantly, one still had no way of knowing what lay in the stretches between genes.

In actuality, the solution of all unresolved questions of molecular biology was blocked by the absence of DNA sequencing methods. As previously mentioned, it was the restriction endonucleases that proved to be the magic wand that moved molecular biology out of its stagnation. They made it possible both to reshuffle genes and to determine nucleotide sequences in DNA. The main stumbling block had been the great length of DNA molecules. Restriction endonucleases made it possible to cut long molecules into sufficiently short pieces. Now only two problems remained: to learn how to separate the pieces and then determine the sequences of each of them.

59

Gel Electrophoresis

A simple physical method called electrophoresis came to the rescue of biologists. A DNA molecule is negatively charged, with the magnitude of the charge being proportional to the chain's length. This is because deoxyribonucleic acid, like any acid, dissociates into a negatively charged moiety and a hydrogen ion. In contrast to a usual acid, however, this occurs in every monomer unit of the DNA polymeric chain. Hydrogen ion dissociates in the phosphate group, shown at the left of the nucleotide depicted in Figure 9. There is no second hydrogen in the phosphate group in DNA, since it separates from the nucleotide during the assembly of the polymer chain. In the process, the adjacent nucleotide loses the OH group of sugar (see the bottom part of the chemical formula in Figure 9). Therefore, the joining of the nucleotide to the end of the DNA polymer chain results in the release of a water molecule.

Each negative charge of the phosphate group of DNA is, of course, matched by a positive ion charge. As a rule, this is an ion of sodium, not one of hydrogen; for although DNA goes under the name of an acid, it is actually a salt. Thus, the famous abbreviation *DNA* with its letter *A* for *acid*, is a misnomer of the first magnitude. (No one, for instance, refers to ordinary salt [NaCl] as hydrochloric acid [HCl]!) However, the name DNA is here to stay.

Rather than sitting on DNA, most positive ions float individually in the solution, forming a rather loose cloud around the molecule. So if a DNA solution is placed in a condenser, the DNA anion will swim toward the positive pole of an electric power supply. The longer the molecule, the bigger the charge, and the bigger the force—and the greater the corresponding increase in the resistance of the medium. If resistance and force grow at different rates depending on length, then speed would also depend on length.

Thus, placing a mixture of DNA fragments of different lengths in an electric field and switching the field off after some time would reveal that the mixture had broken down into several bands, each with molecules of strictly uniform length. This will happen because in the given period of time, fragments of different lengths will have shifted different distances from the point of departure in the field, and those of the same length will have shifted the same distance. However, if we actually try to do this, we will fail to achieve the molecules' separation by their lengths, because in solution the electrostatic force and the resistance depend on the length of the molecule virtually in the same way. This is why the prevailing opinion held that electrophoresis was a worthless technique when it came to DNA separation. A way out of this dilemma, however, was found and would lead to a major methodological breakthrough, resulting in explosive development in both basic research on DNA and its uncountable applications.

We know from our daily experience that some substances, such as gelatins and jellies, although they appear to be liquids, can preserve the shape imparted to them. Similar substances in science have come to be known as gels. What is a gel and what conditions its properties?

A gel is a solution in which polymer molecules are quite heavily entangled and, in some places, tied to one another by chemical bonds. A gel thus represents one single three-dimensional network whose meshes are filled with a solvent. This network acts as a framework that imparts to the whole structure a rigidity quite uncommon for a liquid. The polymer's framework material accounts for a mere few percent of the gel, which consists largely of a solvent. The capacity to assume a gel-like form is one surprising property of macromolecules, that has yet to be comprehensively studied and technically applied.

Living nature makes broad use of gel's remarkable properties. The cornea and the glasslike material that fills the eye's interior are gels. The polymer component consists of proteins; a solvent is, naturally, water. People have long been using gels in food preparation. Gelatins and jellies contain collagen—the protein of connecting tissue—as the polymer component. Another, even more common protein gel is a hard boiled or soft-boiled egg white. Candied fruit jelly is also a gel.

The use of gel as a medium for electrophoresis solved the problem of separation of DNA molecules. Like a snake caught in a fisherman's net, wormlike DNA molecules crawl very slowly to the anode, squeezing through the meshes of the gel network with difficulty. As a result of such snakelike movement, or reptation, the resistance grows with DNA length much quicker than the electric force, and DNA molecules of different lengths are very well separated by electrophoresis in gel. The idea of reptation of polymer molecules in polymer networks, which provides the theoretical background for DNA separation by gel electrophoresis, was put forward by the outstanding French physicist P.-J. de Gennes, who won the Nobel Prize in physics in 1991.

Gel electrophoresis resolution turned out to be so high that DNA fragments of a moderate length differing by as little as one residue could be clearly distinguishable as well-defined bands.

How DNA Texts Are Read

Thus, we can, with the use of restriction endonucleases, cleave DNA into a large number of fragments. Gel electrophoresis makes it possible to isolate every fragment. The procedure is quite simple: When the electric field is turned off, the gel is cut by an ordinary razor blade into pieces, each containing one band or one portion of DNA fragments of strictly uniform length. All that is left is to read the sequence of each fragment. But how does one go about this?

People grappled with this problem for years. Many ideas were tried and discarded. One proposal, for instance, was to bind a compound containing uranium atoms to every nucleotide of a particular type (say, adenosine) and watch through an electron microscope (where, in principle, one can see the heavy atoms) how the markings are distributed along a single DNA strand. The compound containing uranium atoms could then be stitched to thymidine nucleotides, for example. Eventually, the problem was resolved by chemical and biochemical methods. In its final form, the chemical solution was proposed by Harvard scientists A. Maxam and W. Gilbert in 1977. We shall explain the essence of the technique using the example of single-stranded DNA fragments: it is also applicable, however, to double-stranded DNA fragments.

Suppose we have a sample consisting of identical single-stranded DNA fragments of an unknown sequence. First, using a specific enzyme, radioactive phosphorus ^{32}P is attached to one (particular) end of the fragment. (It will be recalled that a single DNA strand has a direction [polarity] in that its ends differ from one another.) We shall refer to the molecule's labeled end as the beginning, with the other end being unlabeled. The sample is then divided into four parts. A substance that breaks the DNA strand after an adenosine nucleotide (A) is added to the first fraction. In so doing, the reaction conditions are chosen so that on the average about one A per fragment will be attacked during the reaction. The reagent will transform the original mixture of uniform-length fragments into one of varying-length fragments. In the process, we are concerned solely with the labeled fragments. If, for example, A nucleotides occupy positions 1, 3, 7, 13, 21, 25, and 26 on the original fragments, we may assume that the reagent is added to result in the appearance of labeled fragments with lengths of 1, 3, 7, 13, 21, 25, and 26 nucleotides. Not a single fragment of a different length is expected to appear.

Similarly, the three other parts of the original solution are treated with substances breaking the chain after T, G, and C. (The attentive reader will note that in our explanation three, not four, different reagents are sufficient.) After this, all four samples are separated in a parallel manner in the same apparatus for gel electrophoresis. After the electric field is cut off, a photographic plate is placed on the gel's surface to imprint on it the labeled bands in the gel (which is why the length of the unlabeled fragments is immaterial). The experiment's result is diagrammed in Figure 16. The sequence can be read directly on the electrophoregram shown to the left. A fragment length that can be deciphered by this method depends on the resolution offered by the gel electrophoresis technique.

In practical terms, this technique makes it possible to sequence fragments with a length up to many hundreds of units. Such a high degree of efficacy was unattainable by previously existing techniques for protein sequencing.

Sequencing all the DNA fragments resulting from the molecule's cleavage appears to be insufficient to ensure the sequencing of the entire DNA molecule. In addition, one also has to know the order in which the fragments are to be joined together. To determine this, it is necessary to cleave DNA once again by using another set of restriction endonucleases and sequence fragments. Sequence superimposition, obtained after different cleavages, permits the establishment of the entire sequence—a task handled by a computer.

Sequencing techniques for DNA have been developed to perfection, based on different conceptual approaches. The most fertile approach proved to be that of Sanger who "taught" the enzymes working on DNA "to read" DNA texts—the same Sanger who learned how to read protein sequences. He is one of the very few who has been awarded two Nobel Prizes.

Sanger's method works in the following way. To start reading a DNA text you first need to know a short sequence, say one that is twenty nucleotides long, that precedes the unknown text. An oligonucleotide complementary to this known sequence is synthesized. This oligomer is called a *primer,* and it serves to initiate synthesis of the complementary DNA strand on the DNA single-stranded template. The synthesis is known as the primer extension reaction. The primer is added to the DNA to be read together with DNA-polymerase. The four deoxyribonucleoside triphosphates (dNTPs:

dATP, dCTP, dGTP, and dTTP) are also added to the mixture because they serve as monomers for the polycondensation reaction of DNA synthesis on the DNA template. Thus, the primer extension reaction starts from one of two primer's termini and proceeds as long as DNA template is available. Normally, the primer extension reaction just leads to the synthesis of double-stranded DNA on a single-stranded template.

Sanger used this well-known primer extension reaction to develop his so-called dideoxy-chain termination method of DNA sequencing. Just as in the Maxam–Gilbert method, the sample is divided into four parts. To each part a small amount of a substance terminating the primer extension reaction is added. These poison molecules are very similar to normal dNTPs: they differ in only one atom in the sugar ring. As we have already learned in Chapter 3, the deoxyribonucleosides differ from the ribonucleosides shown in Figure 6 in the lack of the

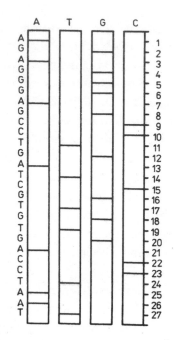

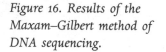

Figure 16. Results of the Maxam–Gilbert method of DNA sequencing.

oxygen atom in the lower right-hand OH group of the sugar. Poisonous dideoxynucleosides are lacking oxygens in both of the OH groups of the sugar. Dideoxynucleoside triphosphates are designated as ddNTPs. When dideoxynucleotide is incorporated into the elongating DNA chain, the further growth of the chain is terminated because the lower left-hand OH group of the sugar is absolutely necessary for addition of the next nucleotide by DNA polymerase. Therefore, when a small amount of ddTTP is added into the first of four parts (remember that T pairs with A), the growing chain is statistically terminated at all positions where adenines are located at the template strand. The higher the ddTTP/dTTP ratio, the higher the probability of chain termination. Similarly, with ddATP, ddCTP, and ddGTP, the chain is statistically terminated at thymines, guanines, and cytosines, respectively. If the other terminus of the primer (opposite to the one initiating the extension reaction) carries the ^{32}P label, after the primer extension reaction is terminated and DNA strands are separated, we will have in each portion a mixture of labeled DNA strands of different lengths, just as in the Maxam–Gilbert method explained above. As a result, after gel electrophoresis, we will obtain a pattern very similar to the one presented in Figure 16.

Nowadays, instead of radiolabeled compounds, the approach uses nucleotide analogues linked to different dyes, which makes sequencing, practiced on an ever-increasing scale, a much cleaner technique ecologically. Special robots have been designed and produced for totally automated reading of DNA texts utilizing the Sanger method. Each day, many thousands of new letters of DNA texts, deciphered by the Sanger method in many laboratories, are added to computer databases. These data are then immediately available to all researchers around the globe via the Internet. Complete DNA texts of many viruses, several bacteria, and the first eukaryote (i.e., an organism whose cells have a definite nucleus)—yeast—are already known. In several years we will know the full DNA text of a human.

The First Surprises

In science, as in any quest, people frequently come up with findings other than those they have sought. Everyone waited for the first sequences to yield telltale evidence on the structure of the regions between genes. There had been numerous conjectures in regard to that structure. The results of the first sequences, however, proved to be disappointing. Indeed, nothing special had been uncovered. The expectation was that by sequencing several promoters recognized by the same RNA polymerase, the mechanism of this recognition would immediately be illuminated. Alas, nothing of the kind happened: Although the sequences turned out to be similar in some ways, it was not clear where the similarity lay.

What was least expected, however, was finding some surprises in the genes themselves. Indeed, the code seemed to have been established once and for all, and it was known for sure that each protein was matched by a particular DNA region—a gene, properly speaking. As a matter of fact, the pious belief in the unassailability of the central dogma of molecular biology had been restored. The shock caused by the discovery of reverse transcriptase had already worn off by the mid-1970s. The first DNA, that of the *E. coli* virus, known by its code name of φX174, had been sequenced. Now, however, came the stunning discovery that the same DNA region encoded information about two proteins!

Could that really be possible? Imagine getting a book with no spaces between the words, only arrows to indicate how the words could be divided—some arrows above and some below the lines. By dividing the text into words using the arrows above the lines you would, say, read *David Copperfield* by Charles Dickens; using the arrows below the lines you would read *The Godfather* by Mario Puzo. "It's just not possible!" you will tell me. And indeed, as far as I know, such a long text just does not exist. But I still remember well from my childhood the following silly jingle in Russian:

NA↑POLE↑ON↓↑KOSIL↓↑TRAVU↓↑POLYA↑KI↓PELI↓↑SOLOVYAMI

Using above-the-line arrows for spaces, the meaning would run as follows: "Napoleon was mowing the grass; the Poles were singing like nightingales"; using the bottom arrows it reads: "In the field he was mowing the grass; the fields were alive with nightingales."

The same situation occurs with φX174 (Figure 17). We can see that the sequence of gene E is wholly locked inside the gene D sequence. Notwithstanding, the sequences of the amino acids of proteins E and D have nothing in common with one another, since they are read with a shifted reading frame. In this respect, the case of the DNA φX174 is much more surprising and captivating than the virtues of the preceding silly jingle. The theoretical possibility of the same DNA region encoding a maximal message of three proteins appears, therefore, to be quite clear. Such an overlapping of three whole genes, albeit on a small stretch, occurs in the G4 phage.

Despite its discovery back in 1977, the gene-overlapping phenomenon has so far failed to provide any explanation as to how a thing like this can occur during the course of evolution. Apart from this surprising phenomenon, however, one can well say that the rest of the sequencing of the first viral DNAs only confirmed the previously established facts. A direct comparison of the sequences of DNA and proteins was carried out to check whether the genetic code had been deciphered correctly. The check revealed that the code had been deciphered without a single mistake.

The thesis about the code's universality was subjected to a crucial new test. Indeed, the very idea underlying genetic engineering—that of the possibility of transposing genes from one organism to another—assumes that the code is universal. It was disclosed that genes transferred to *E. coli* from the most diverse

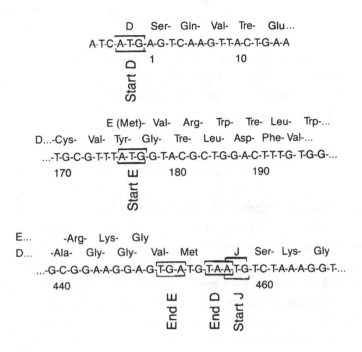

Figure 17. An E. coli (φX174) DNA segment, and the protein chains synthesized on it.

bacteria seemed to perform perfectly there, synthesizing proteins identical to those in the original donor bacteria. RNA obtained from animals, including humans, was used to synthesize a gene with the help of reverse transcriptase. When the gene was then introduced into a bacterium, the protein produced by the latter had an amino acid sequence identical to that of the protein obtained directly from animal cells. What other proof was needed that the code was universal? Then, however, came the discovery that mitochondria have a different code.

The Code of Mitochondria

What sort of creatures are these mitochondria? They are neither bacteria nor viruses, nor are they unicellular. They are just tiny corpuscles swimming in the cytoplasm of cells of eukaryotes. But is it that simple? Not quite.

Mitochondria have a function of great importance to the cell, namely, that of oxidative phosphorylation (i.e., the transmuting of the energy generated by the "burning" of food into energy of adenosine triphosphate [ATP]). In other words, the mitochondrion is the cell's energy-generating plant. Just as electricity is the universal source of energy for household needs, ATP is the universal source of energy to power all enzymes.

ATP is an adenine nucleotide whose phosphate has been joined by another two phosphate groups. By taking away its energy, an enzyme turns ATP into ADP (i.e., adenosine diphosphate) through the removal of one phosphate group. "Recharging," or the rejoining of a phosphate group to ADP, takes place in the mitochondria. This phenomenon, however, has only an incidental bearing on our narrative. We are concerned with a different problem—namely, that mitochondria have their own DNA. Moreover, they possess their own RNA polymerase, which copies RNA from the mitochondrial DNA! Nor is this the whole story. Mitochondria have their own ribosomes and protein-synthesizing machinery. This is quite bizarre, since the very same cytoplasm contains a host of normal cellular ribosomes. However, normal cellular ribosomes use only RNA copies of nuclear DNA as a template for synthesizing protein. For mysterious reasons of their own, mitochondria do not use nuclear mRNAs.

Everything about mitochondria is "undersized"—miniribosomes, mini-RNA polymerase, mini-DNA. This may be very natural, since mitochondria are, in fact, much smaller than cells. However, the capacity to synthesize their own proteins does not at all mean that mitochondria are autonomous parts of cells, independent of nuclear DNA. The size of mitochondria's DNA are so small that they cannot carry all the information of protein molecules needed in order to exist alone. The bulk of the message is inside the cell's nucleus (i.e., written down in the form of a nucleotide sequence in the nuclear DNA). So the set of mitochondria's oddities, it was discovered, was complemented with yet another most surprising peculiarity—the mitochondria have a genetic code of their own.

All this was apparently discovered quite accidentally. B. Barrell and his colleagues from the Laboratory of Molecular Biology in Cambridge (Britain) were attempting to sequence the human mitochondrial DNA. (Incidentally, he was also the first to discover genes' ability to climb one on top of another.) A comparison was made of the sequence of a gene coding one of the subunits of cytochrome oxidase with the amino acid sequence of cytochrome oxidase itself—although that of a bull, not a human. This latter circumstance, however,

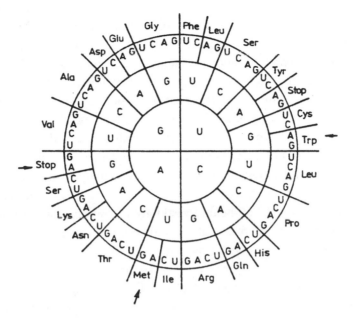

Figure 18. Mitochondrial genetic code. The mitochondria of humans have such a code. The arrows show places where the human mitochondria code differs from the "universal" code shown in Figure 7. In yeast mitochondria (not shown), codons beginning with CU encode threonine, whereas AGA and AGG correspond to arginine.

proved no hindrance in arriving at the precise determination of the genetic code of human mitochondria. The code is depicted in Figure 18.

One can see that the human mitochondria code on the whole looks like the "universal" code already discussed in Chapter 2. However, four codons had changed their meaning: UGA was now the match for tryptophan; AUA was now the match for methionine; and the AGA and AGG codons had become terminating. Miracles, however, did not end there. A comparison of DNA and protein sequences in *yeast* mitochondria produced still another surprise: Their code was different than both the universal and the human mitochondria codes. In addition to the differences found in the code of the human mitochondria, another one was established in the yeast mitochondria— namely, that all four leucine codons beginning with CU had passed over to threonine. Eight codons now showed correspondence to threonine! Leucine had been left with only two—namely, UUA and UUG. In addition, the AUA codon had "come back" to isoleucine, as in the universal code.

How is one to explain these findings? Quite naturally, different inter-pretations may be offered. We can, for instance, say that nothing out of the ordinary has happened. If minor variations had been discovered in the code from the very beginning during the process of deciphering, they would not have been such a surprise. It is true that over the years the attitude toward the code had changed significantly. During the ten years between 1967 and 1977, people had grown accustomed to regarding the code as absolutely universal. Discovery of nonuniversality of the genetic code produced quite a fuss. Indeed, it was no laughing matter to discover in one cell—and a human one at that—the existence of two entire codes. The importance of the discov-ery of new codes should not be underestimated, however, for it was the first tangible evidence that the code had evolved and that it had taken a long time to become what we perceive today.

The view that mitochondria are remnants of a unicellular organism that formed a symbiosis with a eukaryotic cell a very long time ago has been expressed repeatedly. That mitochondria even have a different code is yet another weighty argument in support of such an assumption. It may well be that all cells used to have a code identical to that of the human mitochondria of today, but that the code then changed slightly. It may well be that not all living things on earth owe their origins to cells with the already modified code. Some species may be direct descendants of the ancient cells that had the mitochondrial, "ideal" code. Or there may be species that have evolved from cells that had resulted from other, slight variations of the "ideal" code.

However, a different point of view is possible. According to it, the codes of mitochondria, rather than being more ancient, are actually younger than the basic code, having arisen when a large part of the mitochondrial genes had already passed into the nucleus. The mitochondrial DNA was thus left with so few genes that a code modification ceased to be necessarily lethal to mitochondria and the cell as a whole. A change of this kind, caused by a mutation in the machinery of protein synthesis, triggered mutations in struc-tural genes, which made up for the mutations in the code. Following this, the genes' exodus from mitochondria to the nucleus came to an end, since the mechanism for synthesizing mitochondrial proteins could no longer have its function taken over by the nucleus. The attraction of this hypothesis lies in its offering an explanation of why the transfer of genes from mitochondria to the nucleus had ceased halfway into the process.

CHAPTER 7

Where Do Genes Come From?

∽∽∽∽

The Theory of Evolution and Genetics

The relationship between genetics and the theory of evolution has never been easy. These two sciences use highly reliable, but drastically different research techniques. The evolutionary theory grew out of the study of the enormous variety of the earth's living creatures. Like astronomy, for instance, this science relies on observation. Genetics is a purely experimental science, much akin to physics. (It was no coincidence that Gregor Mendel, founder of genetics, had a solid background in physics, having been a disciple of Christian Doppler.) There is no need to prove that a science relying on observation is, generally speaking, very much inferior to an experimental science in terms of the tempo and possibilities of advancement. One only has to compare the respective accomplishments scored by the theory of evolution and genetics in the twentieth century. In reality, of course, there is no competition between a science of observation and one of experiment, and there can be none. In comparing the two, one should think of a married couple, where differences of opinion and even arguments are common; not of two competing runners.

As genetics continued to amass an impressive record of accomplishments (especially following its passage into the molecular level), its conflict

71

with the theory of evolution, which had arisen at the dawn of this century, became increasingly acute. The substance of the conflict was as follows. The two pillars of the theory of evolution are variability and selection. Presumably, genetics had explained the mechanism of variability—that it was based on point mutations in DNA. But is it precisely the sort of variability that can account for evolution? Perceptive minds understood a long time ago that this kind of variability could hardly get one very far in understanding evolution. All the new things we have learned in the course of the development of molecular genetics have tended to confirm these doubts.

In fact, point mutations lead to individual amino acids being replaced in proteins, in particular enzymes. The word *point* means that the mutation can result in the substitution of only one amino acid residue in one of the proteins of the whole organism. Mutations are rare phenomena, and a simultaneous change of even two amino acid residues in one protein is totally unlikely. But what is the outcome of a single substitution? It will either turn out to be neutral (i.e., not affecting the function of a particular enzyme) or negative.

It is like attaching the tail of a plane to a car. The car will not become airborne, but it will remain roadworthy (only somewhat less so). This is a case of neutral mutation. If you attach the plane's right wing to your car, the car will still fail to get off the ground. Worse still, the fact that the wing knocks against the lampposts along the road will prevent you from riding altogether, and if you try riding on the road's left-hand side, you will end up in an accident. Moreover, attaching a left wing will not get you very far either.

It is clear, therefore, that you need a drastic refitting of the whole of your machine to make the car into a plane. The same is true for a protein. In trying to turn one enzyme into another, point mutations alone would not do the trick. What you would need is a substantial change in the amino acid sequence.

In this situation, rather than being helpful, selection is a major hindrance. One could think, for instance, that by consistently changing amino acids one by one, it will eventually be possible to change the entire sequence substantially and thus the enzyme's spatial structure. These minor changes, however, are bound to eventually result in a situation in which the enzyme has ceased to perform its previous function but has not yet begun its "new duties." It is at this point that it will be destroyed—along with the organism carrying it. The whole thing will have to be started all over again, and with the same chances of success. How can one bridge this hiatus? What can one do to preserve the old function until a new one is well in place?

Classical genetics has failed to offer a model that would allow testing new variants without discarding the old ones altogether. It was this circumstance that spawned a sharp conflict between genetics and the theory of evolution.

Advances made in the study of the genetic organization of bacteria only served to exacerbate the conflict. Through plasmids, bacteria readily exchange the genes they already have. This endows them with a capacity for rapid change. Let us, for instance, take the genes involved with resistance to antibiotics. Far from appearing again and again in each bacterium that has "grown accustomed" to the given antibiotic, as was at one time assumed, these genes get into the bacterium in a "ready-made" form from outside, together with the plasmid.

In addition, one might explain variability of higher organisms on the basis of the regrouping of the "ready-made" genes. But that would mean that genes, having once evolved, are here to stay, with evolution only reshuffling them like a deck of cards. New properties would then be novel combinations of the same old genes. The most unpleasant thing about this scheme is that it creates the impression of explaining the whole set of observations that serve as the basis for the theory of evolution. The century-old experience of breeders is in line with this. All their accomplishments are the result of reshuffling genes previously produced by nature.

Nature itself often uses the same protein design again and again in different organisms and even with totally different purposes. One such example is the protein responsible for vision, *rhodopsin*. Within our eye, the protein absorbs light and sends a signal to our brain. Such signals, received from different rhodopsin molecules located throughout the retina, create the visual image in our brain. Needless to say, rhodopsin from different species that have eyes and brains has the same design. But astonishingly, virtually the same protein (with insignificant amino acid substitutions), dubbed *bacteriorhodopsin*, was found in some bacteria, which obviously have neither eyes nor brains. It also appeared to play a crucial, but totally different, role: Instead of sending signals from eye to brain, bacteriorhodopsin consumes a bacterium with energy being a key protein in the complicated process of conversion of the energy of light into the chemical energy of ATP. The more we learn about genes and their functions in various organisms, the more examples of this sort that are accumulated. However, the key question—of where have the genes actually come from—remains unanswered. Perhaps the bacteriorhodopsin gene first appeared hundreds of million years ago, and much later, Mother Nature, needing a good light antenna to create the eye, adapted bacteriorhodopsin in the form of rhodopsin. Or, on the contrary, perhaps rhodopsin evolved first and then some bacteria adapted the design for their purposes. Darwin's question about the origin of species thus becomes the question about the origin of genes.

Might there be some sort of factory in nature, that manufactures new genes, tests them, and discards the deficient ones? Or might such a gene-producing facility have existed at the early stages of evolution and then, having churned out a huge set of genes, closed down and died a natural death? It would, of course, be much more pleasant if such living gene factories had managed to survive into the present so that lucky researchers could discover them. Perhaps scientists should begin to prepare and equip expeditions to search for these strange relic creatures. Already we have a name for them—the genogens!

Genes in Pieces

But let us make haste slowly. For if the hypothesis put forward by W. Gilbert from Harvard University (who, as you recall from Chapter 6 was awarded a Nobel Prize for his participation in developing the chemical method for DNA sequencing) proves correct, we shall not have to embark on a long journey in search of these bizarre living gene factories, nor shall we have to look for a new name, for "genogens" are nothing but eukaryotes. If that still fails to clarify the situation, then consider this: They are you and me! However, they also include all the other higher organisms: animals, plants, and even some protozoa. Thus, if Gilbert's hypothesis is correct, then there is a wealth of gene factories, and there will be as long as there is life on earth.

Admittedly, this theory owes its origin to a Hobson's choice. Quite simply, it was badly needed to explain some totally unexpected facts that had come to light in the wake of the determination of the very first sequences of DNA of higher organisms, and no other theory existed. Since the amino acid sequence in proteins had been found to be continuous, it was only natural to assume that the nucleotide sequence in genes was also continuous. Numerous experiments on bacteria and bacteriophages have shown this to be the case.

Studying the structure of the genes of higher organisms and their viruses became possible only with the advent of genetic engineering and the development of techniques for DNA sequencing. Now imagine the confusion of biologists faced with the discovery in 1977 that, in higher organisms, the genes, rather than representing continuous sequences, actually consisted of individual fragments separated from one another by other nucleotide sequences! All of a sudden, DNA emerged as a medley of minced genes. When Richard Roberts (then at Cold Spring Harbor Laboratory) and Phillip Sharp (at MIT) independently reported on such a finding, based on the observation of gene organization in a common cold-causing virus called adenovirus, it was greeted as an exceptional case. However, soon afterward it was found that the globin genes of rabbits, the ovalbumin genes of chicken, and the genes of the RNA of the fruit fly *Drosophila* were structured in the same manner. In short, almost all the genes of higher organisms that had been studied turned out to be similarly structured. Eventually, the discovery of Roberts and Sharp became to be considered so important that they were awarded the Nobel Prize in 1993.

The spaces between gene fragments may differ from between a dozen to many thousands of base pairs. How, then, do such "dismembered" genes provide a template for synthesizing whole RNA molecules, which then provide the basis for synthesizing whole protein molecules? It was next found that from a DNA region within which fragments of a particular gene happen to be scattered, a copy is made in the form of a very long RNA molecule. This is a predecessor molecule, or a pro-mRNA. From a pro-mRNA, through a complex process of tailoring (the process is sometimes referred to as "maturation"), "mature" mRNA molecules are obtained that are already in a position to embark on their "immediate duties." In its embryonic (or atavistic) form, the mechanism for mRNA maturation also exists in bacteria, where the function, however, is confined to lopping off the "superfluous " ends in molecules.

How does the maturation process proceed? Specialized enzymes, of course, exist that cleave the pro-mRNA molecule and splice the resulting fragments to one another. But what tells the enzyme how to cleave the molecule correctly and how to splice together the resulting RNA fragments? And how do the in-between spaces get dropped out in the process? The inner workings of such cutting and assembling are far from simple, for if an enzyme just cuts RNA into pieces, Brownian motion will scatter them around, with no hope of Humpty-Dumpty being put back together again!

As has been established, some specialized shorter RNA molecules take part in the process of RNA maturation, or *splicing*, as it has come to be called. These shorter RNA "glue" pro-mRNA in a manner that makes it convenient for specialized enzymes to cut it into pieces and splice those pieces together, getting rid of the superfluous in the process. The DNA fragments whose copy is preserved during splicing are called *exons*, and sections discarded in the process are called *introns*—terms coined by Gilbert.

What advantages, if any, accrue to higher organisms from such an involved mechanism for the production of mRNA? It is not only very complicated, it also can blunder. Indeed, physical and chemical findings indicate that RNA's spatial structure, rather than being rigidly set, fluctuates between widely different states, depending on which sections form the hairpin turns or other elements of the spatial structure. This means that under certain conditions a pro-mRNA will be cut in one manner; under different conditions it will be cut in another. The spaces dropped out will, correspondingly, be different, and the mature mRNA molecules will differ widely. Furthermore, the accumulation of an insignificant number of point mutations (or even just one point mutation) in pro-mRNA may seriously upset the relationship between the spatial structures formed by this molecule.

Gilbert was the first to notice that these deficiencies in the organization of eukaryotic genes, which apparently must make them manifestly inferior to prokaryotes in terms of the accuracy of protein synthesis, may well mean enormous advantages for the process of evolution. Judge for yourself: A great sensitivity to insignificant changes within DNA and the possibility for a simultaneous synthesis of mature mRNA with totally different nucleotide sequences may combine to accomplish the goal—namely, the testing of the most diverse new variants without completely giving up the old ones. This would mean that higher organisms possess precisely the mechanism that was so desperately needed to reconcile genetics with the theory of evolution.

It is held that the exon–intron gene structure was received by eukaryotes from a common ancestor with prokaryotes, a "progenitor" or a hypothetical forefather of all living things on Earth. As a result of evolution, prokaryotes underwent a reduction of their regulatory apparatus and lost the capacity for splicing.

During his study of splicing, Thomas Cech (of the University of Colorado) made a discovery (that earned him a Nobel prize) as stunning in its effect as had been, in its time, the discovery of DNA synthesis on RNA. He found that splicing could occur even without the participation of proteins! RNA cuts itself into pieces without any outside assistance, throws introns out, and "stitches" exons together. True, this "autosplicing," happens rarely, and only with some exotic RNA; but its significance lies in the very possibility of RNA behaving like an enzyme. Before Cech made his discovery, everybody was absolutely convinced that nucleic acids were incapable of anything without the assistance of proteins. RNA molecules with these catalytic functions have received the special name of *ribozymes.*

RNA's capacity to act as an enzyme sheds unexpected light on a central problem of prebiological evolution. Soon after the advent of molecular biology it became clear that biological, Darwinian evolution was bound to have been preceded by an evolution of molecules. But which of the two main biopolymer classes—proteins or nucleic acids—had a more credible claim to preference? Which of them had come about earlier than the other in prebiological evolution? This guessing exercise is very much like the one that has fascinated and kept people busy for centuries, namely, deciding which came first, the chicken or the egg. Indeed, we now know that proteins cannot appear in a cell without DNA and RNA, and DNA and RNA can do nothing without proteins. Still those who tried to unravel the enigma were inclined to take the view that, originally, there were proteins that could reproduce themselves.

The discovery of ribozymes wrought a drastic change in the situation. It now appears most likely that the RNA molecule was the forefather of all living things on earth. RNA's role as the substance of heredity has been known for a long time, from the discovery of RNA-containing viruses. We now know that RNA can act as an enzyme and can apparently catalyze reactions needed for its own reproduction. Only later in the course of further evolution, at the progenitor formation stage, did RNA hand the functions of genetic-information repository over to DNA, which was better suited for the function, and transfer its catalytic functions to protein molecules, with their unique ability to catalyze practically any reaction.

In a modern cell, RNA is assigned a modest role as an ancillary molecule. However, the traces of its former grandeur still abound. In fact, none of the crucial, in-depth processes in the cell can proceed without RNA, even when it seems that RNA could well be done without. For instance, DNA needs a priming for its replication in the form of a shorter RNA. How many RNA molecules take part in protein synthesis? Both mRNA and transfer RNA (tRNA) might well be dispensed with, to say nothing of ribosomal RNA.

For RNA, the forefather, splicing was no big thing. This tends to substantiate Gilbert's theory: Genes' exon–intron structure reflects a very old principle of organization of genetic material. That structure has survived from those antediluvian times when RNA, not DNA, was the most important molecule of the primitive living nature.

Jumping Genes

The advent of genetic engineering and the techniques of DNA sequencing did more than just dispel the notion of genes of higher organisms as continuous DNA sections. It also marked the collapse of the main sacrosanct posture of genetics, which asserted that all cells of the organism have the same set of genes. The validity of this position seemed to have been proven once and for all by Gurdon's experiments, as described at the beginning of Chapter 3. It was discovered, however, that the rule had its own, quite substantial exceptions.

One has to bear in mind that although the "patchwork" structure of genes is the rule for higher organisms, changes in only a few genes in the course of the organism's growth are an exception. What is important, however, is that this exception extends to genes of special importance to the organism: namely, those responsible for immunity.

The capacity for an immune response to an outside invasion is a vital property of the human organism, enabling it to preserve its individuality and to protect itself against alien cells and viruses. One can even say that but for this capacity, people would not be able to live in the present congested conditions. Medieval chronicles are replete with terrifying tales of whole cities and even vast territories decimated by epidemics. Why are we being spared such scourges today? Of course, to a large extent this has been due to improved sanitation and hygiene, but the decisive factor has been vaccines (i.e., inoculation).

Inoculation amounts to a timely warning to the organism of an imminent threat. It switches on the immune system, resulting in an impregnable wall against a potential invader (a bacterium or a virus) before it embarks on its assault. Mass vaccinations have deprived the pathogens of plague, cholera, small pox, and other disease-producing microorganisms—the erstwhile deadly enemies of humankind—of a medium in which they could multiply, thus practically reducing their populations to zero. This is one of the rare cases when the extermination of some of the earth's "creatures" is of no concern whatsoever, even to the most zealous champions of environmental protection.

What enables the immune system to fight various pathogens successfully? The body's weapons for combating diseases are its white blood cells (lymphocytes) and special protein molecules called immunoglobulins (also referred to as antibodies). Lymphocytes form not one but two lines of defense to ward off foreign invaders. The first to recognize and engage the enemy (which may be bacteria, a virus, an alien protein, or some chemical compound; that is some antigen) are T lymphocytes, while B lymphocytes are committed to battle after that. In their membrane, T lymphocytes carry receptors that "recognize" the antigen. B lymphocytes produce antibodies to deal with the intruding antigens. Receptors of T lymphocytes and immunoglobulins are very similar proteins.

The immune system of each organism is capable of generating a vast set of different immunoglobulins. To be more precise, these molecules are similar, being built according to the same master plan, but contain sections called the variable parts, which differ from one another by their amino acid sequences.

Lymphocytes are strictly specialized cells. Each T lymphocyte carries its own receptor, that recognizes a specific antigen.

Thus, the organism is provided well in advance with lymphocytes that can recognize practically any protein, even one that has invaded the organism for the first time. If a virus is a total newcomer in the organism, a T lymphocyte will be found whose receptor will recognize the protein of the virus's particle (which plays the role of an antigen). The binding of the antigen with the receptor sets in motion a very long chain of events. First, so-called T killers, killer cells that destroy the virus-infected cells are produced. Second, one of the B cells, a cell capable of producing the antibody against the receptor-recognized antigen, begins to divide and produce immunoglobulins that bind with virus particles and eventually get them out of the body. It takes time for all this to happen, however. If it is a case of a surprise attack, then before the immune system has had time to respond, the virus will have done a lot of harm or even killed the organism.

It is quite another story if the virus has already "visited" the organism in a noninfectional form at some previous time. Being switched on once, the immune system retains for many years, and sometimes even for life, the capacity to rapidly produce T killers and antibodies against an antigen, should it show up again. These killer cells and antibodies will fall on the virus and nip in the bud its evil designs.

A key question that has long remained unanswered is, "What triggers the organism's reaction to the most diverse antigens?" The fact is that each organism is always ready to produce antibodies in response to practically any foreign protein. At the same time, however, immunoglobulins and T-lymphocyte receptors are quite specific, which means that a molecule, as a rule, recognizes only one specific protein and is baffled by even minimal changes introduced into the protein molecule. Consequently, to combine high specificity with a diversity of immunological reactions, the organism keeps at the ready a host of lymphocytes capable of recognizing practically any antigen.

Are there, then, millions of genes, each of which encodes a receptor and an immunoglobulin of its own? Where do those genes originate? Are they already in the zygote (i.e., inherited from the parents)? Of course they are! How can it be otherwise, since immunoglobulins' chemical structure is determined by the DNA sequence (for what else can determine the structure of proteins)?!

This may well be so, but if our DNA carries messages on the structure of millions of immunoglobulin proteins, how does it have any place left for other messages? We know that besides the immune system, our DNA encodes a great many other things. Also, if we inherit genes of immunoglobulins from our parents, together with genes of other proteins, then why is it that, in our bodies, our mother's immunoglobulins do not attack our father's proteins and vice versa?

Like all people (with the exception of identical, or monozygotic, twins) our parents are immunologically incompatible. One person's immune system attacks the proteins of another. This is the reason why a host of problems occur in organ transplants (e.g., those of a kidney or heart). The fact remains, however, that each one of us produces both proteins inherited from our father and those inherited from our mother, yet nothing terrible occurs. It would terrify one to imagine what would happen if the body produced antibodies to its own proteins. Happily for humans, this happens very rarely.

The paradox, then, lies in the fact that we must explain not why such a disease may strike us, but rather why it does not strike all of us!

In the 1960s, those attempting to explain immunity in genetic terms were distinctly aware of the fact that the very existence of the immune system was clearly in contradiction with the molecular biology of that time. It was clear that there was some mystery behind the whole thing, which, if resolved, could revolutionize our concepts. This is why as soon as a detailed understanding of the genes of higher organisms appeared possible, the genes of immunoglobulins were among the first to be studied. The greatest contribution to resolving the problem through genetic engineering techniques was made by Susumu Tonegawa, who was awarded a Nobel Prize for the following remarkable discoveries.

While he was studying the genes of immunoglobulins at Basel Institute of Immunology at Switzerland in 1976, Tonegawa was actually the first to discover genes in pieces. He found that between DNA sections encoding the variable and constant parts of immunoglobulins, there is one section with no encoded protein sequence. And in a full-fledged immunoglobulin molecule, the variable and constant parts form a single polyamino acid chain. The scientific community learned the news almost at once. Within just a few months it became clear that the "patchwork" pattern was typical of all genes of higher organisms.

However, the news had hardly been accepted and digested when Tonegawa reported a stunning new discovery. He had compared DNA from lymphocytes of an adult mouse with DNA from a mouse embryo. In the embryo, the gene's variable part was found to consist of not one (as in the case of the lymphocytes of the adult mouse) but two fragments marked *J* and *V. J,* the smaller part, is always in the same place, while the longer *V* part is so remote from *J* that it proved impossible to determine the distance to it along DNA.

As in all ordinary cells (not lymphocytes) of an adult organism, the embryo's immunoglobulin genes are structured in a manner diagrammed at the top of Figure 19. There are about 300 *V* genes, four *J* genes, and one *C* gene. The cluster of *V* genes is separated from the cluster of *J* genes by a considerable space. There is also a space separating the *J* genes and the *C* gene, although a much smaller one. Cells with this DNA structure are incapable of producing antibodies. For this reason, an embryo, or even a newborn, lacks antibodies of its own and is equipped with only its mother's antibodies, which had passed into the blood before birth.

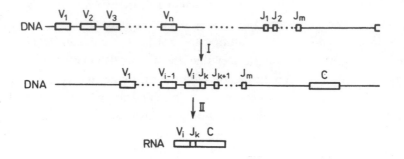

Figure 19. Regrouping of immunoglobulin genes. Stage I occurs during the maturation of lymphocytes. Stage II corresponds to immunoglobulin synthesis in lymphocytes.

The period soon after birth is one in which the organism's own immune system begins to evolve as lymphocytes are formed. Each lymphocyte cell goes through the following stages: A prolonged section is cut out of DNA, beginning at the end of a *V* gene and ending precisely at the start of a *J* gene. The result is a lymphocyte containing DNA structured as diagrammed in the middle part of Figure 19. At the next step, an RNA copy is made from the entire resulting section, which begins with the *V* gene and ends with the *C* gene. The RNA is treated with enzymes in basically the same manner as the process that occurs with any dismembered genes of higher organisms. In the process, RNA is rid of everything, except the copy of the single *VJ* gene that was formed at the previous stage and the *C* gene. All three copies form a single, continuous mRNA chain (bottom diagram, Figure 19) that is translated by ribosomes to produce a protein immunoglobulin chain.

Of course, the most interesting things occur at stage I (i.e., when the lymphocyte of a given type is produced). What determines which pair of *V* and *J* genes will be joined and find themselves side by side when a DNA section is cut out? This is the central issue on which the structure of immunoglobulin, produced by a lymphocyte, is wholly dependent.

The cardinal rule is that all, or almost all, *V* and *J* gene combinations are tried out in the process. This is the first step toward creating a countless variety of immunoglobulins on the basis of a relatively meager set of original genes. With n number of *V* genes and m number of *J* genes we can get nm number of different pairs. Therefore, when $n = 300$ and $m = 4$, as was

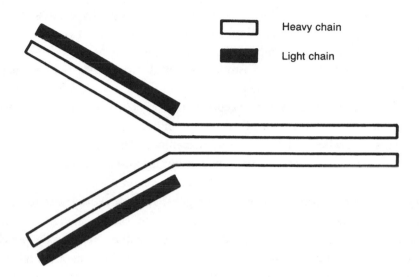

Figure 20. Structure of an immunoglobulin molecule.

mentioned earlier, the resulting number of different antibodies will be in the vicinity of 1,000. This, however, is not the whole story. The immunoglobulin molecule consists of four, rather than one, polyamino acid chains—two light and two heavy ones (Figure 20). Like the two heavy chains, the two light ones are identical. However, both the heavy and the light chains are synthesized independently, and the reshuffling discussed above occurs in both. That is why 1,000 has to be raised to the power of 2. Thus we arrive at 1 million different antibodies.

The story is still not complete, however, for it was discovered that at a certain point in time a mechanism that affects mutations is switched on, but only in the V genes. The DNA regions around the V genes remain unchanged and random nucleotide substitutions occur in the V genes. This further multiplies the variety of immunoglobulins many times over.

We have examined events occurring in B lymphocytes. Exactly the same process takes place in T lymphocytes, resulting in the formation of a host of receptors.

One can only marvel at nature's ingenuity in devising a mechanism that supplies us with lymphocytes and the antibodies that they churn out that are able to handle every conceivable emergency. A riddle that had tenaciously resisted the efforts of several generations of physicians and biologists rapidly yielded to a combined onslaught by specialists armed with the know-how for manipulating the DNA molecule.

The vista that unfolded astounded even the molecular biologists accustomed to the sensations of the last decades. In fact, it was no joking matter to discover that gene reshuffling occurs in the organism of humans, of every mammal, and broader still, of every vertebrate; inducing the formation of millions of new genes, a process that is accompanied by intensive mutagenesis.

Scientists had thought (no, they were absolutely convinced) that the blueprint for the organism's structure was to remain unchanged as soon as a zygote had been formed. They had thought that the blueprint was absolutely identical in all cells and that it was only a matter of cells "reading" their own, different parts of the blueprint. Of course, no one had ever denied the possibility of gene mutations during the organism's development, although they tended to be explained away as accidental hindrances or errors in the planned development of the organism. Now it turned out, however, that during the course of its development, every organism evolves its own, absolutely unique set of genes (immunoglobulin genes at any rate, and maybe some others as well). This is one of the factors that determines the individuality or the "ego," of every vertebrate.

There is no doubt that, as with all previous outstanding scientific advances, the discovery of gene restructuring during the course of the organism's individual development will have a significant impact on our notions of ourselves and our world.

As a science, molecular immunology, now experiencing a boom caused by the invasion of genetic engineering methods, still has a host of unsolved problems. Why is an organism smart enough to know not to produce antibodies against its own proteins? Some mechanism of rejecting certain lymphocytes probably exists. One gargantuan task is to find a link between immunity and cancer, for the fact is that, like immunity, the disease is the lot of vertebrates. Between these two crucial phenomena in the life of humans, immunity and cancer, there undoubtedly exists a very close link. The link is being studied at all levels, including that of DNA. One can say with certainty that the discovery of the possibility of gene regroupment during the course of development sheds a totally new light on the problem of cancer. We shall return to these questions in Chapter 12.

CHAPTER 8

Circular DNA

∞∞∞

DNA Rings

The attentive reader must have noticed that to understand DNA's biological functions, scientists rely only on the information that the molecule consists of two complementary strands and that the genetic message is encoded in a sequence of four types of nucleotides (A, T, G and C). These two facts serve as the foundation for the whole stately edifice of modern molecular biology, including genetic engineering and biotechnology. Even the fact that DNA is a helix, rather than a simple ladder, let alone the finer details of the molecule's physical structure, now are viewed as superfluous pieces of knowledge by some biologists, and even more so by some genetic engineers. In fact, the adherents to this view say that it is high time we stopped rummaging in DNA and got down to dealing with only purely practical tasks that the present state of knowledge is capable of resolving.

This position is clearly shortsighted. The material presented in previous chapters provides us with convincing proof that in the study of DNA, even the most seemingly insignificant facts may lead to discoveries of paramount importance.

The history of the discovery of restriction endonucleases is just one case in point. It all started with an attempt to clarify the methylation of a negligible quantity of nucleotides, a very fine chemical characteristic of the DNA molecule. This characteristic was not associated with the principal DNA functions, and was believed to only permit the cell to tell its own molecules from "alien" ones. Where would molecular biology, genetic engineering, and biotechnology be now if the restriction enzymes had not been discovered?

Who would be bold enough to assert that a meticulous study of DNA's structure would not enlighten us about some totally new characteristics of the molecule that are important for its functioning, and would not reveal the existence of new, previously undetected enzymes? Who could discount the likelihood that, as a result, we would gain the ability to manipulate genetic processes still more actively? The studies of recent years seem to inspire in us the confidence that it is precisely the eventual understanding of the biological role of the fine characteristics of DNA's structure, and the structure of DNA in complexes with proteins, that promises the most interesting and unexpected findings. The discovery of circular DNA, the phenomenon of supercoiling, and the topoisomerase enzymes may be the most palpable proof thereof. Clarification of the questions that arose in the process of these discoveries forced molecular biologists to rely heavily on assistance from physicists and mathematicians.

When biochemists learned to isolate DNA molecules from cells (and they mastered the art a very long time ago), they soon learned that the molecules behaved like common linear polymers. There were two ends per molecule. Nobody seemed to doubt that all the DNA molecules were linear chains. However, geneticists were frequently in the dark as to which genes were to be regarded as the end ones. Consequently, they were forced to draw their genetic maps in the form of circular diagrams. However, one can imagine the Homeric laughter that would have been triggered by a crank who dared to assert that such assumed circular maps reflected the genuine circular structure of the real molecules! To make an assertion like this in earnest, one needed proof that DNA molecules indeed can be circular. And as is frequently the case, the answer came from totally unexpected quarters.

Electron microscopists were studying small DNA of oncogenic (i.e., cancer-inducing) viruses. Genetic information on these DNA was all but nonexistent, but they were convenient to work with: They did not break into pieces, as is often the case with long molecules, which are extremely difficult to isolate in an intact form. To their immense surprise, the microscopists discovered in the early 1960s that some of the viral DNA were in the shape of closed rings. It now became clear that the circular genetic maps had by no means been accidental.

This discovery, however, failed to attract any special interest. DNA can probably assume many forms in a virus. Sometimes it has the form of only one of two complementary strands, and sometimes the strand forms a closed ring. Moreover, as we already know, some viruses carry the blueprint of their development in the form of RNA, which is converted into DNA only within the cell by a special enzyme, reverse transcriptase. It is also common knowledge that in many cases, DNA has a regular linear form inside a viral particle. However, the search for circular DNA went on. Gradually it was discovered that in many cases, even if DNA was linear in a viral particle, it closed into a ring following the virus's penetration of the cell. It was found that prior to the start of replication, such a linear molecule assumes a form (referred to as replicative) in which the two complementary strands of the DNA form closed rings (Figure 21). DNA of bacteria, including those of *E. coli*, were found to be circular. Plasmids, the gene carriers par excellence in genetic engineering, are always circular. In brief, it is difficult to name cases of DNA working in a cell in which it does not assume a circular shape.

Why should a cell wish to see DNA molecules closed into rings? What advantages are inherent in this shape? What changes does this induce in the molecule's properties? To answer these questions, one has to take a closer look at this new DNA form.

Supercoiling and Topoisomerases

The matter of utmost importance at present is that in a DNA molecule, the complementary strands twine around one another like two lianas, and once each of the strands closes, the two resulting rings cannot be separated; they form a link. The simplest link—the symbol of wedlock (Figure 22)—is known to everybody.

The quantitative degree of the linkage of two rings is described by an Lk value referred to as the linking number. It is very easy to determine the value for any link. One only needs to imagine that a soap film is spread over one ring and then to calculate the number of times the second ring pierces the film. One will see easily that for the wedlock symbol, $Lk = 1$; however, for the link diagrammed in Figure 21, $Lk = 9$.

The Lk value is remarkable for its constancy in a given ring pair, no matter how hard we try to deform these rings (although not to the snapping

Figure 21. In a circular closed DNA, two complementary strands form linkage of a higher order.

point, of course). This provides mathematicians with grounds to say that the *Lk* value is the topological invariant of a system consisting of a pair of rings. Without a hand from mathematicians, molecular biologists would never have understood the properties of circular DNA.

Thus, if we have turned DNA into a circular, closed molecule, then the linking number of its two strands will not be changed by whatever we do with the molecule, as long as the sugar–phosphate chains forming the backbone of the respective complementary strands remain intact. It is this circumstance that gives closed circular (cc) DNA special properties that sharply distinguish them from linear molecules; principally, that ccDNA may be endowed with surplus energy in the form of so-called supercoils.

To explain what we have just said, let us imagine linear DNA placed in some definite ambient conditions. In DNA of this kind, there is a definite number of base pairs per one turn of the double helix. This is the γ_0 value. In the Watson–Crick double helix, $\gamma_0 = 10$, but it can exhibit slight variations (of mere tenths of fractions, which, however, is of importance to us now) in a changing environment. Let us further assume that the linear molecule has been turned into a circular one through a

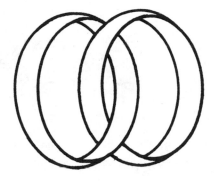

Figure 22. The simplest linkage—a symbol of conjugal union.

minimum of force. The simplest thing to imagine is that we have turned the molecule into a circle and "glued together" the respective ends of each of the two strands. What will the *Lk* value be? Clearly, $Lk = N/\gamma_0$, where N is the number of base pairs in the molecule.

Now let us change the ambient conditions. The DNA molecule will acquire another balanced value of the number of base pairs per turn γ'_0. What happens then? The molecule strives to gain the corresponding linking number value, which now should be $Lk' = N/\gamma'_0$, but cannot come into its own, since a different *Lk* value has already been imposed on it.

This can also happen in marriage unions! When the marriage alliance was concluded, the *Lk* values equaled 1. However, if the conditions change, one of the parties may wish to dissolve the marriage (i.e., to make *Lk* equal 0), and a very tense situation may develop in the process. Something akin to this happens to DNA as well. The molecule finds itself in a tense, and energywise, nonlucrative, state of supercoiling.

Ordinarily, supercoiled molecules assume the form diagrammed in Figure 23. Quantitatively, supercoiling or superhelicity is described by the value $\tau = Lk - N/\gamma'_0$. Just as the double helix is ascribed a definite sign (positive for the right-handed helix and negative for the left-handed), the superhelix, in the same vein, can, in principle, be either positive or negative. In Figure 23 the double helix is right-handed, as it should be for DNA, and the supercoiling is negative.

This latter assertion may appear confusing, for it seems that the superhelix in Figure 23 is right-, not left-handed. This is one of the paradoxes encountered in

Figure 23. The shape that supercoiled DNA usually assumes. The supercoiling is left-handed, and therefore negative.

the study of supercoiling. To gain a better understanding of what this is all about, take a piece of rubber hose about three feet long, and preferably quite rigid. Insert into one end of the hose a pintle with a part of it sticking out, and turn the hose into a ring by inserting the remainder of the pintle into the other end of the hose. The important thing is to prevent the hose ends from turning freely with respect to one another following the linkage.

You can now proceed to model supercoiling: Holding one end still, turn the hose's other end around the pintle axis to make the hose axis form a left-handed helix. After this, allow the closed-ring-shaped hose to assume a form that suits it best, by holding it with two fingers of one hand. You will see that it will assume a form similar to the one diagrammed in Figure 23.

The painstaking isolation of newer DNA from cells and the determination of their state increasingly provided investigators with evidence that in addition to being closed into rings, DNA were supercoiled. Furthermore, the supercoiling in all cases was found to be negative.

It was now clear that DNA's "supercoiled" state was the rule rather than the exception, as had been believed previously. At the same time, however, the doubt arose as to whether DNA actually had this form inside the cell. This forced the admission that most likely this was not the case. Supercoiling appeared to be DNA's reaction to its forcible removal from its native environment, for the conditions in which DNA exists inside the cell naturally differ from its environment following its removal.

Inside the cell, DNA is bound with some proteins, including those that open up the double helix and unwind the two strands in those places. Because of the unwinding, however, the average γ_0 value for the molecule becomes bigger than for pure DNA unassociated with proteins. For this reason, even if DNA is not twisted into a superhelix inside the cell, ridding it of its proteins will inescapably cause its transformation into a superhelical state with a negative sign.

This was the simplest explanation for DNA's supercoiling, as accepted in the early 1970s. It meant that supercoiling had no biological significance whatsoever.

At this time, the problem of DNA's supercoiling was being studied by only two teams of researchers. One was led by Jerome Vinograd (Caltech), who discovered the phenomenon of supercoiling, the other by James Wang (Harvard). What would prompt investigators to study a DNA property that was clearly of no biological significance? In truth, Wang joined in the effort only because he was determined to clarify the ability of particular proteins to unwind DNA.

Wang's experiments were time- and labor-consuming: He nicked one strand of ccDNA, created a complex between protein and the nicked DNA, then cured the nick with a DNA ligase, separated DNA from the protein, and, finally, measured the value of supercoiling. It would be ideal to have one protein that would both nick the strand and cure the rupture, Wang reasoned. This would save so much time. Wang set out to look for such a protein in cell extracts of *E. coli*.

What could assist him in the search? The properties sought were clear: If the necessary protein did exist, then it would assist the supercoiled DNA in transforming into a closed circular molecule with no supercoils. In fact, as soon as a protein has nicked one of the strands, the tension in the DNA will immediately disappear (i.e., the superhelix will be no more). However, the curing of the nick by the protein would result in a DNA in which $Lk = N/\gamma_o$. In other words, the hunting of the enzyme capable of changing the Lk value had begun in earnest.

Wang managed to find such an enzyme. The protein turned out to be the father of a vast group of enzymes later baptized as topoisomerases for their ability to modify DNA's topological properties. (Wang originally called the enzyme he found ω-protein; now it is called topoisomerase I). The discovery of topoisomerases forced many to doubt that supercoiling was totally useless in the biological realm. For if enzymes with a capability for changing topology actually existed, this meant that the cell could not be totally indifferent to topology.

Thus, a planned quest for topoisomerases began in earnest. In 1976, the team of Martin Gellert (National Institutes of Health) discovered an enzyme that, assisted by ATP (a universal battery accumulating energy in the cell), had the opposite effect of the one produced by the protein discovered by Wang. The enzyme, called DNA gyrase, transforms the relaxed, non-supercoiled ccDNA into a superhelix. At this point, however, it was found that if DNA gyrase was damaged, then the cell's most vital processes, including DNA replication, ground to a complete halt. It thus became clear that supercoiling was the state of DNA vital to the cell.

Why Supercoiled?

Supercoiling is one of the most important examples of how the physical state of the DNA molecule affects its behavior in the cell. This problem is the object of intensive study by specialists in fields ranging from medicine to mathematics. It is therefore not surprising to observe a great variety of hypotheses about the role of supercoiling in the functioning of the cell. We shall dwell in more detail on one of these, which at present appears to be the simplest and the most plausible.

This hypothesis owes its origins to the unquestionable proof that, to begin replicating, the DNA molecule first has to twine itself into a superhelix, which, however, is totally unnecessary for the replication process itself. Moreover, one of the strands of the closed circular DNA is sometimes nicked before replication has begun, the agent of the rupture being a special protein that comes into play only if DNA is supercoiled. Thus, a kind of a catch-22 situation develops: The cell tries to supercoil with the help of one protein (DNA gyrase) so that another protein immediately liquidates that supercoiling. However, the irrefutable fact remains that replication will not begin without this enigmatic rite, at least as regards the objects that have been studied so far (e.g., the ϕX174 bacteriophage).

Only one explanation appears to exist. The ritual described earlier is nothing more than the checking of DNA for the integrity of the sugar–phosphate chain, or a kind of quality control system for DNA. Indeed, one should not forget that DNA in the cell is constantly being damaged by irradiation, chemical agents, its own nucleases, and thermal motion for that matter. The cell possesses a whole paraphernalia of means, referred to as the repair system, to mend the damages. In Chapter 3 we spoke of how the repair system mends damage done by UV radiation. The system has a host of enzymes at its disposal. Some, namely nucleases, nick the DNA strand in the vicinity of the damaged nucleotide. Others move in to widen the gap and remove the damaged section. However, the genetic message is preserved in the process, because of the existence of the second, complementary strand from which Kornberg's DNA polymerase (DNA polymerase I) restores the severed chain.

Thus, a continuous process of repairing the wound inflicted on the DNA molecule occurs in the cell. The universal and unfailing remedy is surgical interference involving nicking one of the strands of the double helix. What will happen if the time of replication happens to coincide with that of repair? The DNA polymerase engaged in replication will halt upon reaching the nick: Neither process will be able to proceed further, which will amount to a catastrophe. This means that replication ought to be started only upon obtaining hard evidence that the repair has been completed (i.e., we must be convinced that both strands are intact).

But how can one verify that? You could "dispatch" some protein on its way along DNA in order to "sound" the molecule. However, other proteins might be "sitting" on the DNA, which would stop the "trial-balloon" protein. In addition, this sort of verification would take a long time indeed. Who could guarantee that no new damage would be done while the integrity of the chain was being checked in this manner, section by section? No, this would not work.

It is here that supercoiling comes to the rescue. The fact is that supercoiling is only possible in DNA in which both strands are intact throughout. It is quite simple to find out whether DNA is supercoiled, for in a supercoiled DNA, it is much easier to separate the two complementary strands (i.e., to open up a double-helix section). The opening up is akin to the action exerted by the unwinding protein: It eases the strain in negatively supercoiled DNA. Thus, the protein, whose function is one of control, will come into contact with the given DNA section (it will recognize it by its definite nucleotide sequence) and try to separate the strands in that particular place. If the bid is successful, replication begins from that spot in no time at all. If, however, it has proved impossible to separate the strands, then one will have to wait, since DNA is thus not yet ready for replication.

Does this sound very much like how we check an electric cable? Rather than feeling with our fingers the whole length of it, we simply pass electric current through it. If the current passes, everything is all right; if not, we start looking for a faulty connection. Upon finding and eliminating the defect, we again pass the electric current to make sure that there are no other ruptures. In any case, no one would put an electric cable to use without first checking it. However, DNA is not a conductor and no current passes through it. Consequently, the cell had to invent its own tester—a rather smart one, we must admit.

It has been proven that supercoiling has other functions besides ensuring the start of replication. To understand the relationship between DNA supercoiling and transcription, try conducting the following experiment: Go to the window and rotate the double string used to draw the blind in a clockwise manner, long enough to obtain a double helix. Then insert a pen between the two strands and push it ahead without rotation. You would be making a mechanistic model of the transcription process: The pen, or RNA polymerase, moves along DNA, the interwound double string. It can be seen from this experiment that while RNA polymerase translocates, it should overwind DNA in front of itself and underwind DNA behind itself. In other words, DNA becomes positively supercoiled in front of RNA polymerase and negatively supercoiled behind it. Although an inevitable consequence of the DNA helical structure, this wave of supercoiling that moves with RNA polymerase seems fantastic. However, the elegant experiments by James Wang and co-workers at Harvard demonstrated that these waves of supercoiling actually exist both in prokaryotic and eukaryotic cells.

If you continue the experiment with the string and pen, you will find that very soon the pen stops, because the double string cannot be overwound any more. Thus, one has to suppose either that DNA and RNA polymerase rotate around each other or that the cell can eliminate both positive and negative supercoils. One can hardly expect that a very long DNA molecule and a very bulky transcription machinery loaded in prokaryotes with even bulkier translation machinery could rotate around each other. On the other hand, topoisomerases are known to be able to change DNA supercoiling. Based on this simple fact, Leroy Liu and James Wang put forth the concept of waves of supercoiling. But how can the wave of supercoiling be measured? Extracting DNA from the cell and ridding it of proteins causes the memory of the wave to be lost, as the wave of supercoiling does not change the overall DNA linking number.

Although Wang and co-workers could not directly observe the wave of supercoiling in vivo, they unambiguously demonstrated its existence using the inhibition of different DNA topoisomerases. Their most striking piece of evidence is the formation of positively supercoiled plasmid DNA in *E. coli* when DNA gyrase is inhibited. In this case, topoisomerase I continues to remove negative supercoils, while positive supercoils, which are normally removed by DNA gyrase, are accumulated in DNA.

These findings throw a totally new light on the biological significance of DNA supercoiling. Specifically, people had thought that DNA gyrase in *E. coli* served as an enzyme that introduced negative supercoils into DNA. Competing with topoisomerase I, it supported the native level of DNA supercoiling in the cell. Some held that by changing DNA supercoiling, gene expression could be regulated. Now it can be seen that this picture, which was almost generally accepted, is really upside down. Indeed, DNA gyrase seems to eliminate positive supercoils rather than to create negative ones in *E. coli*. Native supercoiling is an irrelevant notion because actual local supercoiling may be highly positive, highly negative, or negligible; depending on the position of promoters, on the current position of RNA polymerase, and on the relationship between the rate of RNA polymerase translocation and the efficiency of supercoil removal by the topoisomerases. Supercoiling depends on transcription to a far greater extent than transcription depends on supercoiling.

Physicists and Mathematicians at Work

To gain a proper understanding of the role played by supercoiling, one has to study comprehensively its impact both on DNA's biological functions and on its physical structure. This challenge was met by physicists: Their efforts, however, were plagued by serious problems from the outset. The different physical methods tried in measuring the value of supercoiling tended to yield diverging results.

Vinograd, the discoverer of DNA supercoiling, met Brock Fuller, a mathematician from Caltech, and asked for his help in understanding the problem of circular DNA. Fuller took a keen interest in what Vinograd shared with him, and felt that mathematical findings concerning the unexpected link discovered between topology and differential geometry could be of use in tackling the problem.

These two fields of mathematics study similar subjects—curves and surfaces—but from totally different perspectives. Differential geometry studies local properties of the surface, such as curvature and torque. Topology, on the other hand, is indifferent to these characteristics; it is concerned with properties such as whether the surface has holes (the shape of the holes being immaterial), how many there are, and so on. Thus, a marble statue may be studied by both a geologist and an art scholar. The geologist will be interested only in the stone, and the art scholar will be interested in the shape imparted to the stone by the artist. Approaching the problem from their strictly professional angles, the two would hardly find a common language.

The link between the differential-geometrical and topological characteristics of one class of surfaces—namely, bilateral ribbons—was yet another surprise for mathematicians. The famous Möbius strip is one of the ribbon varieties. To make it, take a strip of paper, twist it once, and then glue the two ends together. Now select a starting point at random and draw a pencil line parallel to the strip's edges. You will soon find yourself returning to the point of departure. And if you look at the strip carefully, you will find that every site on the strip carries your pencil's mark. This is the remarkable, even somewhat enigmatic property of the Möbius strip, which happens to have only one side and hence is referred to as a unilateral ribbon.

Now cut another paper strip and glue its ends again. This time however, twist the ends not by $180°$, as was the case with the Möbius strip, but at an angle equal to m $360°$, where m is an integer. You will always get a bilateral ribbon. With a bilateral ribbon, both edges are closed curves that either are not linked with each other or form a linkage with some value of the linking number Lk, it being clear that $Lk = m$.

Fuller understood at once that from the mathematical viewpoint, a ccDNA molecule represented a bilateral ribbon with the sugar–phosphate backbones of the molecule serving as the edges of the ribbon. Consequently, the fact that ccDNA can only be a bilateral ribbon is a purely chemical fact, having to do with the existence of directionality in each of the DNA strands, with the complementary strands having opposite directions. If one tried to form a Möbius strip out of such a molecule, one would surely fail, since the ends of the complementary strands would reach one another "head to head" or "tail to tail" and fail to connect.

What Fuller reported in his article soon after his conversation with Vinograd can be summarized as follows. The topological characteristic of ccDNA— the Lk value—cannot be expressed unambiguously in terms of a single geometrical and, thus, physical characteristic of the molecule. It is expressed through two geometrical characteristics. One characteristic is well known in differential geometry: the twisting number, Tw—the axial twisting of the ribbon. This is the number of turns completed by a vector, which lies on the ribbon's surface perpendicular to its axis, during the course of its motion along the ribbon. The second characteristic originally had no name, but Fuller gave it the name of the writhing number, Wr (and so the vogue of giving exotic names, which first arose in high-energy physics—just recall the quarks, charm, the color, and so on—was gradually infecting mathematicians as well).

The findings cited by Fuller were proven with scientific strictness by James White from UCLA in 1968. White found a clear, precise relation between Lk, Tw, and Wr:

$$Lk = Tw + Wr$$

What is amazing about all of this is that such a simple equation was discovered so late. It proved to be of inestimable value in the study of the properties of circular DNA. So what, then, is the significance of this equation, whose simplicity seems to border on primitiveness?

One major consideration is that the Wr value depends only on the shape the ribbon's axis has in space and is totally independent of how the ribbon has been twisted around its axis. Furthermore, there exists for the Wr value a general formula that makes it possible to calculate the value for any curve.

Another, totally unorthodox feature is that on the left side of White's equation there is a value that can only be an integer (this is inherent in the definition of Lk, since the number of piercings of the imaginary film mentioned earlier can only be an integer). On the other hand, terms on the right can assume any value and need not necessarily be integers.

At this point in our narration, an avalanche of perplexing questions may arise. The Tw value is the number of times the ribbon twines around its axis. Why then is the result not an integer if the ribbon is closed? Does any writhing number, in fact, occur? What is the difference between Lk and Tw? Do we not, in calculating Lk and Tw arrive, although by different routes, at one and the same value?

Let us stage an experiment to set the record straight. Cut out a narrow strip of paper with a width of half an inch and wind it around a finger (or a cylinder as in Figure 24) several times. Then force the ends of the strip to

protrude slightly so you can glue them together. This will not result in a significant
axial twist. In this way a closed ribbon is
obtained with $Tw = 0$, because of the
method used. But what will the Lk value be?
This can also be found experimentally.
Pierce the ribbon with scissors in any randomly selected place and cut the ribbon
along its entire length. The result will be
two very narrow interwound strips. Their
linking number will be precisely the Lk
value for the edges of the original ribbon.

Thus, it turns out that an Lk value can
be created without creating any Tw. What
was done in winding the ribbon around the
cylinder was to impart a writhing to it. The
equality of $Lk = Tw$ holds true for all cases
when the ribbon's axis lies on the plane.
The feeling that this must always be the
case is due to the fact that we ordinarily
tend to think of the ribbon's (or a DNA molecule's) axis as a simple figure— say, a circle
or something of that sort.

After the appearance of Fuller's paper
it became clear that the contradictions that
surfaced in the study of supercoiling owed
their origins to the fact that some methods
measured the physical characteristics that
depend on Wr and others measured those
that depend on Tw. Equipped with reliable
mathematical tools, physicists embarked on
a planned study of supercoiling's influence
on the properties of ccDNA.

Gaining popularity at precisely the
same time was the method of gel electrophoresis, which, as discussed in Chapter 6,
was found to have a very high resolution in

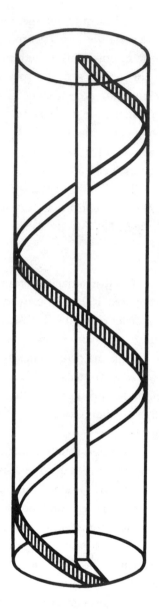

*Figure 24. A ribbon wound
around a cylinder.*

separating DNA molecules of different lengths. A German scientist, Walter Keller, came up with the "crazy" idea of trying to use gel electrophoresis to separate ccDNA molecules with different *Lk* values. The principle for separation, however, had to be totally different from the one employed for linear DNA molecules of different lengths.

In this case, the length of molecules, differing only in the number of supercoils, is the same. Consequently, both the charge and the force operating from the electric field would also be the same. However, the speed of a molecule's motion in gel is determined not only by the force applied to the molecule, but also by the resistance it generates during its motion. This, in turn, depends on the molecule's shape. It is clear that a molecule with the shape of a severely twisted rope, as diagrammed in Figure 23, will meet with much less resistance from the medium while moving in the electric field than will an unfolded molecule. In other words, the greater the absolute value of the writhing number, the faster the molecule's motion. We are speaking of the writhing number rather than *Lk*, since the medium's resistance, determined by the spatial shape of the double-helix axis, is virtually independent of how the helix is twisted around the axis.

Following this line of reasoning, Keller proceeded to experiment with gel electrophoresis and soon was able to demonstrate that putting a sample of a superhelical DNA isolated from a cell on a gel would result in a set of individual bands separated from one another by about equal distances. The *Lk* value is the only discrete characteristic of ccDNA. This means that DNA molecules in these bands may differ only by the *Lk* value. The most plausible assumption was that the neighboring bands would differ in *Lk* value by the figure of 1. This was later proven to be the case.

The result of separating ccDNA molecules that differ in the number of supercoils is shown in Figure 25. On the right-hand side is a photo of gel following the completion of electrophoresis. To make DNA visible, the gel is stained with a fluorescent dye that binds firmly with DNA and "labels" it. On the left-hand side is a graph showing dependence of the intensity of the dye's fluorescence on the gel coordinate. One can see the high degree of clear-cut separation that can be achieved. Using such figures, it is not difficult to calculate the value of supercoiling for each band.

Gel electrophoresis has been as productive in the study of circular DNA and supercoiling as in DNA sequencing. Many fine measurements have been performed that have made it possible to determine the major characteristics

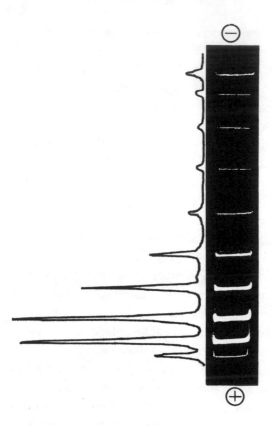

Figure 25. Separation of DNA molecules that differ in the number of superhelical turns, using the gel electrophoresis technique. The experiment was conducted with DNA from a small pAo3 plasmid, containing 1,683 nucleotide pairs. Originally the molecules were added from the top, near the negative electrode (not shown).

of ccDNA. It was the method of gel electrophoresis that permitted a precise measurement of the energy that can be accumulated inside DNA with the help of supercoiling.

What changes can be wrought by supercoiling in DNA's structure? It stands to reason that one would describe any structural change that would result in easing the tension caused in ccDNA by supercoiling as advantageous. It was clear, therefore, that supercoiling must contribute to the formation in the double helix of open regions and *cruciform structures*. Cruciform structures may appear in DNA in invertedly repeated sequences—palindromes.

Palindromes exist in any language, not just that of DNA. They are words, verses, or sentences that read the same forward and backward. Some are short, like *radar* or *Hannah*, but some are quite long. Take, for instance, this Latin one about moths: *In girum imus noctes et consumimur igni.* When you read the phrase from left to right then from right to left, you will find that the pronunciation and the meaning will be the same (the spaces between the words and the punctuation marks being disregarded).

Palindromes are frequently encountered in DNA texts. Since DNA consists of two strands (i.e., as if they were two parallel texts), its palindromes may be of two types. Palindromes in an ordinary, single text are called "mirrorlike." But the more frequently encountered palindromes in DNA are ones that read alike along either strand in the direction determined by the chemical structure of DNA. (It will be recalled that the two complementary DNA strands have opposite directions.)

In practically all cases, the sections recognized by restriction endonucleases are palindromes. What follows are a few examples: On the left are the names of restriction endonucleases (the names are quite quaint, since they include the first three letters of the Latin name of the bacterium from which the given restriction enzyme has been isolated); the arrows show the sites where DNA has been cut by a restriction enzyme.

	—G$^|$AATTC→
Eco RI	
	←CTTAA$_|$G—
	—CCC$^|$GGG→
Sma I	
	←GGG$_|$CCC—
	—CTGCA$^|$G→
Pst I	
	←G$_|$ACGTC—

The remarkable feature of DNA palindromes is their ability to form cruciform structures. In fact, considering that the left half of the palindrome is bound to be complementary to the right one, the following scheme will hold true for the place of recognition by the Eco RI restriction endonuclease and, in the same vein, for any other palindrome:

$$\cap$$
$$AT$$
$$AT$$
$$-GC-$$
$$-CG-$$
$$TA$$
$$TA$$
$$\cup$$

In any case, this does not contradict the complementarity principle. It does, however, immediately raise a number of questions. Is the DNA backbone capable of making the U-turns that are bound to arise at the top and bottom of the cruciform? Since a DNA strand possesses a certain rigidity, it is not a simple thing for it to make a U-turn. We have already discussed in Chapter 3 the problem in connection with the difficulties involved in DNA being housed in chromosomes. The double helix is quite rigid, and, to aid its bending, there exist in the chromosomes specialized proteins such as histones and others. It is true that a single strand is much less rigid; thus, U-turns in a single strand are generally possible. But they are energetically unfavorable. Thus, it is not at all clear why a cruciform structure should arise in DNA if it can easily transform itself into a regular double helix. All this, however, is true solely with respect to linear molecules. But what about supercoiled molecules?

In general, the formation of a cruciform structure removes strain. Can this, then, benefit supercoiled DNA? What degree of supercoiling is required for this advantageous transformation? To answer these questions, a team of biophysicists from the Institute of Molecular Genetics in Moscow—which included Vadim Anshelevich, Alex Vologodskii, Alex Lukashin, and myself—conducted in 1979 a detailed analysis of the formation of the open regions and cruciform structures in linear and supercoiled DNA. The theoretical analysis revealed that the formation of both open regions and cruciform structures in linear DNA is very unlikely. The probability is especially small for the cruciform structure, being on the order of 10^{-15} (i.e., being practically zero).

The situation changes drastically with the increase in negative supercoiling. The probability of cruciform formation in short palindromes, like the ones recognized by restriction enzymes, remains negligible in any degree of supercoiling. However, the longer palindromes, with fifteen to twenty or more pairs, present quite another picture. Although rare, such palindromes do crop up in sequenced DNA. One example, for instance, is the palindrome from a ColEl plasmid, shown in Figure 26. According to our calculations, in long palindromes the probability of cruciform formation increases sharply with the increase in superhelicity. With normal superhelicity values typical of many DNA, the likelihood of the formation of a cruciform turns out to be on the order of unity—that is, 10^{15} (a million billions!) times more than in the case of a linear molecule. After our theoretical predictions had been published, many experimentalists began looking for cruciform structures in ccDNA. David Lilley (Dundee University, Great Britain) and Robert Wells (University of Alabama at Birmingham) managed to beat all others to the goal by proving that palindromes in natural supercoiled DNA do form cruciform structures.

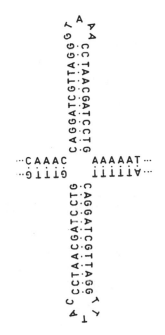

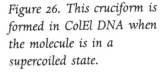

Figure 26. This cruciform is formed in ColEl DNA when the molecule is in a supercoiled state.

How did this prove possible? The fact is that the hairpins themselves, arising in cruciform structures, are too small to be seen even under an electron microscope. This is why a different technique was used in the search for cruciforms. A superhelical DNA was treated with an enzyme—a single-strand-specific endonuclease. The enzyme nicks only a single DNA strand and "gives the double helix a wide berth." The enzyme thus does not cleave an ordinary linear or circular closed DNA molecule that is not supercoiled. It was discovered, however, that it does cleave a supercoiled DNA, but in a definite site. The researchers studied nucleotide sequences to the right and to the left of the site of the cleavage. In all cases, the cleavage occurred strictly in the middle of large palindromes, in precisely those that, according to theoretical calculations, had to have the ability to form cruciforms. Such results seemed to point to one explanation: In the sites where long palindromes occur in supercoiled DNA, it is very probable that the double helix will turn into a cruciform structure and that in the process, a single-strand-specific endonuclease will cleave the single-stranded loops at the top and bottom of the cruciform.

Soon, however, a question was raised. Are cruciforms actually formed in the DNA, or do they arise only under the effect of protein, the single-strand-specific endonuclease? To answer this question, it was necessary to register the formation of cruciforms by a different technique, one not involving the use of enzymes. As we have already said, cruciforms cannot be seen directly, even under an electron microscope. So what were we to do? Once again, the gel electrophoresis technique came in quite handy.

Indeed, the services this delightfully simple technique has continued to render to DNA scientists are truly priceless. We have accepted the fact that scientific progress is unthinkable without sophisticated and costly installations stuffed with ultramodern electronics, powerful computers, lasers, and God knows what. Such installations are developed over many years by giant firms employing many thousands of people, and cost scores or even hundreds of thousands of dollars. If you find yourself in a laboratory whose DNA research is widely known and ask to be shown the experimental facilities at which the research is conducted, you will be amazed.

They will take you to a room with nothing in it except an ordinary chemical bench. Amid all the vials with reagents, they will show you a small, clearly homemade, transparent Plexiglas box. The box is partly filled with water and has two thin wires sticking out. That's about it. They will also tell you that an important experiment is in progress. You will feel like you are in the presence of a magician and are making a fool of yourself. "You just can't be serious," you will exclaim, "pretending with this primitive thing to investigate the most complex problems of DNA structure, for even the most powerful microscopes and other wonders of state-of-the-art technology have been tried and found wanting! This is just a hoax."

It is not a hoax. It is just that, as with any good magician's performance, you will miss the essence of the trick. The magic box would indeed be utterly irrelevant, if it did not contain a transparent gel plate, superimposed on which is a DNA specimen that, of course, you will fail to see. The most important point is what type of DNA it is. The molecule has been precisely prepared with the help of the most sophisticated techniques of genetic engineering. That is to say, before being put into that simple box, the DNA has passed through the hands of many people, the greatest experts in the field working in laboratories scattered around the world. Every one of them has used his or her entire repertoire of expertise to modify the DNA's properties in the manner required. Then, at the final stage, in one of the rooms nearby, the molecule undergoes the last preparations for the planned experiment. So in reality, what appears to be primitive and simple, is anything but. Study of the structure of DNA is a complex process; one that in recent years would not be possible without the close cooperation that has occurred between DNA structure specialists and genetic engineers.

But let us return to cruciforms. The use of the gel electrophoresis technique for registering cruciforms in DNA is based on the fact that transition of a DNA section with a palindrome sequence into a cruciform results in a partial removal of the superhelical stresses in the molecule. The molecule straightens out and begins moving slowly in gel under the influence of the electric field because of the strong resistance of the environment. As a result, the topoisomer in which a cruciform has formed turns out, as in the electrophoregram in Figure 25, to be higher than the topoisomer that has one negative superhelical turn less but carries no cruciform. Consequently, the appearance of cruciforms is bound to upset the regular "ladder," in which topoisomers with a growing negative supercoiling move faster and faster. What happens is superimposition of two "ladders" corresponding to topoisomers with cruciforms and those without them. The result of the experiment is a complex pattern of bands that is difficult to decipher.

Of help in understanding the pattern is a smart technique that was invented in the early 1980s and baptized as two-dimensional gel electrophoresis. The experiment is conducted not with a gel column, as is the case with the traditional, one-dimensional gel electrophoresis, but with a quadrangular gel plate. A DNA specimen is placed on one angle of the plate and an electric field is applied to two opposite sides of the plate. The result is a pattern of bands arranged along the side of the plate. The picture is identical to that resulting from one-dimensional gel electrophoresis. Next, the electrodes are reswitched, so that the electric field becomes perpendicular to the field in which the first separation was done. In the process, gel is saturated with chloroquine molecules, which bind with the DNA double helix and decrease the axial twist (namely, the Tw value). The binding results in a sharp reduction of superhelical tension in all the specimen's topoisomers, so that the stress is no longer sufficient to cause the formation of cruciforms. Thus, the cruciforms disappear. Consequently, gel electrophoresis in the second direction is bound to result in only one, regular "ladder." The end result is shown in Figure 27.

The most important thing in the figure is that a discontinuity occurs in the regular pattern of spots. With the help of a single-strand-specific endonuclease, it is easy to see that a cruciform has actually formed in all topoisomers coming after the discontinuity. Two-dimensional gel electrophoresis conducted after treating the DNA preparation with a single-strand-specific endonuclease will result in the disappearance from the pattern of all spots coming after the discontinuity. This is because endonuclease attacks and undoes the single-strand loops of the cruciform. DNA loses its closed pattern, topological stresses are removed, and all topoisomers turn into either straightened rings or (following a prolonged treatment with single-strand-specific endonuclease) linear molecules.

Further examination to localize the rupture spot on the DNA molecule, reveals that the endonuclease "inflicts" the rupture precisely in the center of the principal palindrome. Such experiments, first conducted in 1983 by Victor Lyamichev and Igor Panyutin in our laboratory (at the Institute of Molecular Genetics at Moscow), have finally demonstrated beyond a doubt that cruciforms do indeed appear spontaneously in DNA with sufficient negative supercoiling. The experiments have also shown our theoretical predictions of the probability of cruciform formation in supercoiled DNA to have been correct in quantitative terms.

What is the role of cruciforms in DNA? Nothing is known as yet about this. It is surmised that cruciform structures may serve as "landing areas" for some proteins on DNA. In any case, cruciforms represent the first reliably proven example of how the structure of individual sections of a biologically active DNA may be significantly modified under conditions close to those in which DNA has to function in a living cell. Clarification of the importance of the role in DNA's functioning in the cell played by cruciforms and other irregularities (covered in chapter 11), is a topic for further research.

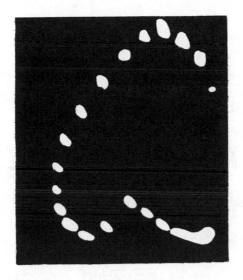

Figure 27. A typical pattern of two-dimensional gel electrophoresis, observed during the formation in DNA of cruciform or other alternative structures (which will be discussed in Chapter 11). A specially prepared mixture of different topoisomers of the same DNA, carrying an insert capable of changing into an alternative structure, was placed in the left top angle of a quadrangular gel plate. Then an electric field was applied to force DNA molecules to move from top to bottom along the left edge of the plate. Following the separation of topoisomers in the first direction, the gel was saturated with chloroquine molecules, which lessen superhelical stress. Chloroquine concentration was chosen so as to make superhelical stress insufficient for the formation of an alternative structure. Then the direction of the electric field was changed to force molecules to move from left to right. As a result, the sequence of spots in the second direction corresponded to the topoisomers' sequence.

The uppermost spot corresponds to zero topoisomers (i.e., to a relaxed and nonsupercoiled DNA). The spots clockwise to that spot correspond to positive topoisomers, and those going counterclockwise, correspond to negative ones. One can clearly see the mobility drop, observed in this case between -10 and -12 topoisomers. This means that in topoisomers -12, -13, . . . , an alternative structure is present, whereas in topoisomers . . . , -9, -10 it is absent. Topoisomer -11 occupies an intermediate position: in it the alternative structure now forms, now disappears.

CHAPTER 9

Knotted DNA

ooooo

About Knots

Everybody knows what a knot is. We tie many knots every day. The simplest knot looks something like this:

$$(1)$$

Now how about this one:

$$(2)$$

Pausing to think, the reader will comment, "Well, it is just a ring twisted into a plait. This second example has nothing to do with knots, and does not belong in the same category." Although in appearance, the plait does not seem to be a knot, the fact is that, like the ring, of which the plait is made,

$$(3)$$

it may well have a more valid claim to the title of a knot than the first figure. A mathematician will describe the second or the third figures as trivial knots and will dismiss the first as not being a knot at all.

"Oh, these mathematicians!" you probably think. "They will always get you into a mess." I might agree with that. I am no mathematician and often grumble in a similar fashion myself. In this particular case, however, I will beg to differ with you.

You can, of course, call the first figure a knot, but then try to explain what makes it different from this one:

$$\rule{3cm}{1.2pt}$$

(4)

The fact is that the first figure can always be disentangled, making the chain a straight line. This would be impossible to achieve if the ends of the chain happen to be infinitely long. It is much better to get rid of the ends altogether:

(5)

Now just try to disentangle this one! The difference between the third and the fifth figures is clear to everyone: You cannot transform one into the other without breaking the chain. The knot in the fifth figure is referred to as a trefoil, or a clover leaf, since it can be transformed to look like this:

(6)

I think that you will now agree that the notion of a knot, in the strict sense of the word, applies only to closed chains (although in everyday life it is common to refer to objects like those shown in the first figure as knots).

We already know of two types of knots: the trivial one (whose status among knots is similar to that of zero among numbers) and the trefoil (figure 5 or 6). After the trefoil, the next knot in terms of complexity is a knot called a figure eight knot. It looks like this:

(7)

How about this one?

(8)

Imagine that this entangled thing is a rope. Can you, without cutting the rope, transform it into a simple ring (a trivial knot), into a trefoil, or into a figure eight knot? Or is it impossible? In other words, what is the simplest shape into which you can disentangle this knot?

The first to take a serious interest in knots was the British physicist and mathematician P. Tait in the 1860s. Physicists at that time (as they do now) wished to understand the structure of the elementary particles of matter, and thought that particles could have the form of vortexes of electricity. In a letter to Tait, James Clark Maxwell (a famous British physicist and the author of Maxwell equations on electrodynamics) wrote, "What if the vortex gets knotted?"—and he drew a trefoil.

Tait had a knack for abstract mathematical constructions. He started thinking about what other kinds of knots exist. Soon he forgot about particles altogether (as though he knew for sure that in more than 100 years they would still be a problem) and began to spend hours on end making all sorts of knots with a rope. It was Tait who compiled the first table of knots, consisting of a whole sequence of the knots he had been able to devise. Eventually, a complete inventory was taken of all possible knots with less than ten crossings on their projections. There proved to be eighty-four such knots. A number of them are depicted in Figure 28.

Tait recruited mathematicians to solve the problem of the knots. Having worked with the riddles of knots for about sixty years, mathematicians had become singularly proficient in disentangling very complicated ones. By 1928, they had devised a good knot invariant.

The knot invariant is an algebraic expression whose value remains the same no matter how you try to entangle the knot, provided you do not rupture the rope. In principle, the ability to calculate the invariant permits the disentanglement of the knot. It is only necessary to determine the invariant of a given knot and then to compare it with the values of the invariants calculated for the knots included in the table. Although mathematicians have recently come out with a series of new knot invariants (they haven't stopped working on the knot problem since Tait's time), the so-called Alexander polynomial $\Delta(t)$ still remains the most handy one. For the trivial knot, $\Delta(t) = 1$. For the trefoil, $\Delta(t) = t^2 - t + 1$. For an eight, $\Delta(t) = t^2 - 3t + 1$, and so on. Rather than being characterized by a number, each knot is thus characterized by a whole algebraic expression that has a certain variable t devoid of any special sense.

If you have mastered the Alexander polynomial, it will not take you long to see that the knot in the eighth figure on p. 109 is essentially a trivial, albeit quite an entangled one.

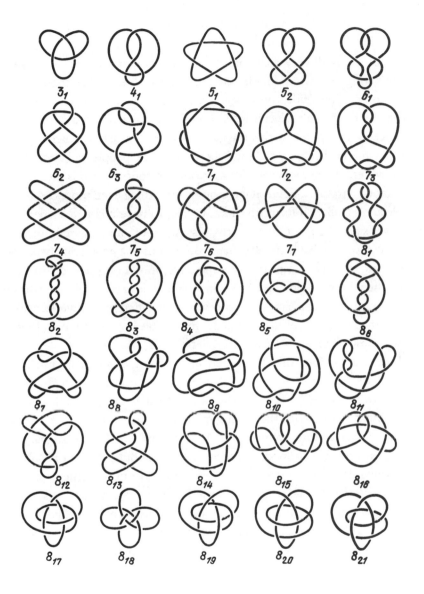

Figure 28. Examples of knots with less than ten crossings on their projections. The knots are arranged according to the growing minimal number of crossings on their projections. For a trefoil (3_1), there are three; for an eight, four. Different knots with the same number of crossings are grouped together in the table, with each knot being marked with an extra index beside the indication of the number of crossings.

Knots in Polymers

"This, of course, is all very well and good," you may say, "and, moreover, quite captivating. But what does it all have to do with the DNA molecule?" In fact, I do beg your indulgence; I got somewhat carried away.

The idea of tying a molecule into a knot began to be debated seriously in the early 1960s. It may well be that the idea had been canvassed earlier as well, but only tongue in cheek. By the early 1960s, however, people had arrived on the scene who no longer saw the idea as funny. Of course, what is meant is a genuine knot—a trefoil, a figure eight knot, or something even more complex. The molecule's ability to form trivial knots (i.e., to have a closed shape) has been known since the great German chemist A. Von Kekulé figured out that the benzene molecule has a circular shape. But just have a go at tying benzene into a knot! You will find this totally impossible, for its ring has too small a hole. And then, how would you tie it? A molecule is not a piece of rope whose ends you can tie by hand. You can, however, try a different technique: Make the ends of the molecule cohesive, with the reasonable hope that the molecule may form a knot if its ends chance to meet.

Thus, it is obvious that to be tied into a knot, a molecule must be sufficiently long. But how long? More generally, the question, which is not easy to answer, can be formulated this way: What is the probability for a nontrivial knot to arise with the closing of a chain consisting of n segments?

Rather than talking about atoms or residues, we are talking about segments, because it is more reasonable to speak about a certain idealized chain in which the word *segment* (or, rather, Kuhn statistical segment) stands for a more or less rectilinear section. In rigid polymer chains, such a section includes numerous atoms and even many residues. The so-called freely joined chain with which theoreticians model real polymer molecules (although this model, like any, has a limited sphere of application, which the patient reader will have the chance to see) looks roughly like this:

$$\text{} \qquad (9)$$

Does this seem familiar? This is precisely the "drunkard's walk" discussed in Chapter 3. The above figure is a planar analog of an unclosed polymer chain of ten segments. The ringlets between the segments stand for "hinges."

Now imagine yourself forcing this freely joined chain into an accidental closing in three-dimensional space; this happens many times over. (No
knots whatsoever are possible on a plane. Interestingly, there are no knots in
four-dimensional space either. Knots are possible only in three-dimensional
space.)

How many nontrivial knots will result in the process? Their fraction
among all closed chains will be precisely the measure of the probability of
the knot formation. Do not, however, try to guess that probability. The
rabbit will stay in the hat! Intuition will be of no use.

About twenty-five years ago my co-workers—Anshelevich, Lukashin,
and Vologodskii—and I became obsessed with this problem. At that time, we
knew nothing as yet about Tait, his table of knots, or the existence of Alexander polynomials. We wasted hours on end discussing possible ways to assess
this probability. For instance, we discussed quite seriously a project involving
the construction, out of some material, of a large cubic (or some other)
lattice. We would then take a rope and insert it along the ribs of the lattice.
The direction the rope would follow at each node of the lattice was to be
decided by casting a die. It was possible to invent a procedure whereby you
could generate only closed chains. We then could connect the rope's ends,
remove it from the lattice, and disentangle it to find out the type of the
resulting knot.

The only thing that deterred us from implementing our project was
that we did not know of a convenient technique for removing the rope from
the lattice. Now we know, however, that even had we obviated that difficulty (e.g., by using a collapsible lattice), we would spend the rest of our days
messing around with this foolish construction, mocked by our own performance. A bit later I shall tell you why.

Fortunately, at a very handy time indeed, we hit upon a Russian edition
of the book *Introduction to Knot Theory* (Ginn and Company, 1963), written by
two American mathematicians, R. Crowell and R. Fox. From this book we
learned about Alexander polynomials, the table of knots, and many other
things. Then it became clear to us how we should proceed. Instead of tying
knots ourselves, we left it to a computer. We found it possible, among other
things, to teach the computer to calculate the Alexander polynomials and
thus to disentangle knots.

And what did we get in the way of results? It was discovered that the
probability of knot formation depended on more than just the number of

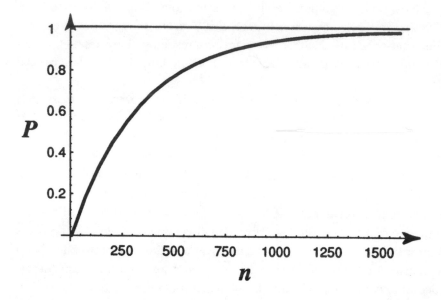

Figure 29. The probability (P) of formation of knots in a polymer dependent on the number (n) of segments in it. The curve was obtained from computer calculations.

segments n in a chain (which is the reason why I have tried to dissuade you from engaging in guesswork). If a chain is very flexible (i.e., if each segment contains a very small number of atoms), then the probability of the formation of a knot is infinitesimally small. Even when $n = 100$, one will observe only 1 knot per 10,000 cases. So, as you can see, when I was talking of spending the rest of our lives participating in an inglorious exercise in futility, they were not empty words. We thus became aware of the reason for the dismal failure that awaits all attempts to synthesize a knot by the technique proposed by chemists—that of inducing an accidental closing of simple polymer hydrocarbon chains: These chains are too flexible to allow for the practical possibility of knot formation.

It is quite a different matter when it comes to very rigid chains whose segments contain many residues. The probability of knot formation in such chains is much higher. Figure 29 gives the results of the calculations of this probability. One can see that for comparatively short polymer molecules, the probability of formation of a nontrivial knot increases almost linearly with the growth in the number of segments, approaching 0.5 when $n = 200$. As the chains get longer, the probability of knot formation levels off asymptotically approaching 1. For short polymer molecules, for which the probability is low, the bottom line is that the more complicated the knot, the less chance there is for its formation. This can be seen from the following diagram showing the distribution of different knots (in percentages):

n	3_1	4_1	5_1	5_2	All others
10	98	2	0	0	0
20	83	10	1	3	3
40	76	12	5	3	4
60	66	13	4	6	11
80	58	13	6	6	17
100	58	14	3	6	19
120	58	9	2	6	25
140	53	8	2	9	28
160	46	10	5	8	31

So far it has proved impossible to synthesize a knot by purely chemical means. The German chemist H. Schill, one of the fathers of catenanes directional synthesis, has been trying this for many years already. A catenane looks like this:

(10)

or like this:

(11)

It looks like a pocket watch chain, doesn't it? Mathematicians refer to such structures as *links*.

Knots in Single-Stranded DNA

During the year following the publication of our calculations on the probability of knot formation in polymer chains, it indeed proved possible to tie DNA into a knot. Wang and his co-workers treated single-stranded DNA rings with topoisomerase I (which they had discovered), and then put the preparation under an electron microscope. Of course, using a microscope you cannot tell genuine knots from entangled trivial ones. However, the authors of the experiment asserted that under the same conditions, the original molecules that had not been treated with topoisomerase formed unfolded rings with practically no crossings.

These findings, as well as other arguments that we shall omit here for brevity's sake, leave no doubt that Wang and his co-workers actually managed to tie a single-stranded DNA molecule into a knot. These molecules constitute the bulk of the preparation—approximately 90 percent. "But wait a minute," you will say, "these findings are totally at variance with the theoretical calculations presented on the previous page!" In fact, one could hardly expect the high efficiency of knot formation in single-stranded DNA, as represented in these calculations.

Wang gave a trenchant explanation for this contradiction. In his opinion, given the conditions under which the experiment proceeded, one should in no way liken a single-stranded DNA to a simple freely joined chain, as was the case in the preceding calculations. Any nucleotide sequence of sufficient length always possesses complementary sections that, finding each other, form short helices.

Of course, the whole issue is not confined to helical sections. Single-stranded DNA has a knack for assuming a highly whimsical spatial configuration. In the process, the closing of the chain into a ring tends to generate inevitable strain that would disappear if the ring is cut. Topoisomerase most likely binds with helical sections and nicks one strand, after which one part of the molecule may then begin to rotate freely around the other. This results in relaxation, or the easing of the strain. Topoisomerase then proceeds to seal the nick, thus fixing the new state of the molecule. Et voilà—you have a knot!

Knots in Double-Stranded DNA

Thus, the molecule of DNA has been tied into a knot for the first time. Molecular biologists succeeded where synthetical chemists had failed. This, of course, was only the beginning. Another tempting goal ahead was that of tying a double-stranded DNA molecule into a knot. In principle, this is not difficult to do. DNA of the bacteriophage lambda appeared to be the most eligible candidate for the experiment.

This phage was the most fertile object of a study by the Phage Group that was gathered by Delbrück in his time. Together with *E. coli*, its host cell, it became the chief testing "site" in the study of gene replication and transcription.

Although linear in a phage particle, DNA of phage lambda has "cohesive" ends—single-stranded, mutually complementary sections of 12 nucleotides each, looking roughly like this:

(12)

If such a DNA is isolated in its pure form (it is commercially available) and given a chance to float freely in a solution, the cohesive ends would close together, thus forming a circular DNA. Given the lambda DNA molecule's considerable length (around 50,000 nucleotide pairs), there is a high degree of probability that, having formed a ring, the molecule will be tied into a knot.

It will be recalled that a double-stranded DNA is a very rigid chain, with segments containing about 300 base pairs. This is why the diagram in Figure 29, based on computer calculations, may be used for assessing the probability of knot formation in a double-stranded DNA. One can deduce from the diagram that, when closed into rings, about half of the phage lambda DNA molecules will form knots. The hitch, however, is that in the case of such a long molecule, it is difficult to tell a nontrivial knot from a trivial one. Thus far, any attempts to draw such a distinction have failed. In actuality, lambda DNA has cohesive ends precisely so that it may assume a circular form once it enters into the host cell. If DNA does not close into a ring, it will not be able to replicate itself or function normally (if, of course, the devastation it causes once inside *E. coli* can be described as normal functioning).

Here arises a question that needs to be answered. What will happen if, upon forming a ring, DNA does get tied into a knot? Theory tells us that this is a distinct possibility. Will this not hinder the viral DNA's performance in the cell? The viral DNA is expected to produce a host of copies of itself. However, if getting tied into a knot prevents this, then the cell must have mechanisms for preventing knot formation. But what are they?

You can easily convince yourself that DNA that is tied into a knot will have difficulty replicating itself by trying this experiement: Take a strip of paper and make a nontrivial knot out of it—a trefoil, for instance—by gluing its ends together. Then use scissors to cut the strip lengthwise into two parts. This would demonstrate at least one of the possible ways of DNA replication. You will find it impossible to separate the two resulting knots. My colleagues and I raised these questions in 1975 in the paper in which we reported our first calculations of the probability of knot formation. The answers came five years later.

In 1980, Leroy Liu, Chung-Cheng Liu, and Bruce Alberts from the University of California at San Francisco reported that their many years of research had finally resulted in identifying the set of conditions under which a regular double-stranded DNA formed knots. The researchers had worked with short, circular molecules, rather than with long DNA, whose knots are difficult to discover because of their great length. They found that the addition of a lavish portion of topoisomerase of a particular type stimulated very active knot formation. One can judge the formation of knots by the appearance of highly mobile DNA fractions during gel electrophoresis. During the next step, they proceeded to treat knotted molecules with a low concentration of topoisomerase mixed with ATP. Lo and behold, the knots came untied!

This latter development fully corresponded to the theory, since the DNA used was short and the equilibrium fraction of knots in it was not to have exceeded 5 percent.

Knot formation in the presence of an excessive quantity of the enzyme has probably been due to the fact that upon binding with DNA, the protein modifies the molecule's physical properties by enhancing the adhesion of regions remote from one another along the chain. Results of calculations demonstrate that this sticking together sharply increases the probability of knot formation.

Discovery by Alberts and his co-workers of a new, knotted form of duplex DNA spawned a flurry of similar publications. Methods of genetic engineering were immediately applied in attempts to form knots. It was clear that there were two types of topoisomerases. One type made knots in single-stranded DNA and came to be called topoisomerases I; the other, "specialized" in double-stranded molecules, and was baptized topoisomerases II. This, however, was not the whole story. It was also discovered that topoisomerases II (which also included DNA gyrase), besides their ability to tie and untie knots, could combine two or more DNA molecules into catenanes (i.e., interlock them like links in a chain).

The discovery of proteins' ability to tie knots in DNA generated keen interest. It helped to shed light on the workings of topoisomerases, including DNA gyrase—a key enzyme of the whole class. In fact, you cannot tie a circular, closed DNA into a knot without breaking the double helix. Just breaking the chain, however, is not enough. You also have to pull the molecule's other part through the resulting "peephole" and then seal the latter. Topoisomerase II happens to be capable of performing this complex job! Using X-ray crystallography, Wang and his co-workers obtained remarkable pictures of topoisomerase II. These pictures have made it possible to reconstruct all stages of the enzyme action process.

In the presence of topoisomerase II, DNA behaves as if the ban on material bodies passing through each other does not apply to it. All this, of course, is due to the presence of the enzyme, for without it nothing of this kind would ever happen, since DNA is not an electron or alpha particle for which the effect of quantum tunneling is possible. Topoisomerases allow DNA to behave in the same bizarre manner. It would be as if, while playing tennis, you sent the ball into the net, saw it pass through the net without obstruction, and then found the net totally intact. Thus, the mechanism by which the cell resolves DNA's topological problems, including the one of replication of knotted molecules, became clear.

Does the preceding also explain the fact that DNA gyrase changes supercoiling? To be sure, it does. One can see from Figure 30 that the pulling of one DNA section through another results in the appearance of a superhelix (a) because of the change in the Wr value by two. This is precisely what has been found experimentally. Unlike topoisomerases I, which change the DNA Lk value by any integer, topoisomerases II change Lk only by even numbers.

Type I topoisomerases also follow the same modus operandi of causing breaks and pulling the strand through the resulting "peephole." The only difference seems to be that, unlike topoisomerases II, the type I topoisomerases perform the trick with a single- rather than a double-stranded DNA. Thus, it is likely that knots in a single-stranded DNA are tied by topoisomerase I in the same way that topoisomerase II ties them in a double-stranded molecule.

The discovery of topoisomerases and the understanding of their underlying mechanism undercut one major objection to the theory of the double helix, repeatedly upheld or rebutted over the past forty plus years. Many people

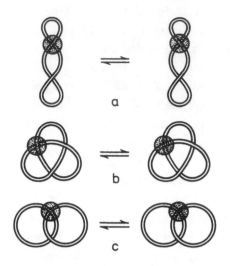

Figure 30. Three "topological reactions," catalyzed by a type II topoisomerase: (a) change in the number of superhelical turns ($\Delta Lk = \pm 2$); (b) undoing and tying of knots; (c) splitting and formation of catenanes.

during this period were puzzled by the fact that DNA has to unwind in replication. Does it really spin around in the cell like the cable of a speedometer?

People held widely divergent views of this phenomenon. Some saw nothing puzzling about it. Others tended to dismiss the problem, saying that it would work itself out with time. Still others were busy inventing ingenious explanations. One theoretical physicist asserted, for instance, that one strand could pass through the other by means of quantum tunneling. Finally, there were those who saw this phenomenon as a clear defect of the Watson–Crick model. They insisted that at least in the cell, DNA was not a double helix.

The fence sitters' patience has been rewarded. It is quite likely that topoisomerases resolve all topological problems faced by the DNA double helix. At any rate, they are capable of creating in the cell conditions under which strands can "tunnel" through one another. How this actually occurs in the cell has yet to be clarified, but one thing is quite clear: The chief argument used by the critics of the double helix for many years has been debunked.

Thus, persistent attempts to tie DNA into a knot have unexpectedly led to a resolution of the old controversy about the double helix. Still the question remained whether our predictions about the DNA knotting probability were quantitatively correct. Two groups, S. Shaw and J. Wang at Harvard and V. Rybenkov, N. Cozzarelli, and A. Vologodskii at Berkeley, have recently arrived at an unambiguous answer to this question by studying cyclization of DNAs with cohesive ends. The DNAs they studied were much shorter base pairs (10,000) than lambda DNA, and for them various types of knots differed in their mobility in gel. This made it possible to measure experimentally the fraction of knots of different types in the sample of DNA molecules, which became circular due to the cohesive ends. The fraction of knots gave the probability of knot formation. Quantitative data on the knotting probability obtained independently by the two groups proved to be in excellent agreement with our theoretical predictions.

Until now, knots could be formed only by using artificial techniques. However, one can assume the existence in nature of a new, knotted form of DNA. Reports have already appeared to the effect that DNA is knotted in the "heads" of certain mutant bacteriophages. But can there be knotted DNA under normal, natural conditions? The answer to this question has yet to be found.

CHAPTER 10

Genetic Engineering: Hazards and Hopes

∞∞∞

The century of biology upon which we are now well embarked is no matter of trivialities. It is a movement of really heroic dimensions, one of the great episodes in man's intellectual history. The scientists who are carrying the movement forward talk in terms of nucleo-proteins, of ultra-centrifuges, of biochemical genetics, of electrophoresis, of the electron microscope, of molecular morphology, of radioactive isotopes. But do not be fooled into thinking this is more gadgetry. This is the dependable way to seek a solution of the cancer and polio problems, the problem of rheumatism and of the heart. This is the knowledge on which we must base our solution of the population and food problems. This is the understanding of life.

—W. WEAVER, 1949

Science and Invention

Over the entire period of its existence as a species on the planet Earth, Homo sapiens has not undergone any significant change in the biological sense. Children born now are the perfect images of children delivered by our female ancestors 10,000 years ago. But how different the world has become! The globe is covered by a thick web of railways and highways, and thousands of aircraft and spaceships follow their invisible routes in the atmosphere and beyond. Man has visited the Moon, and man-made vehicles have been to Mars and Venus, have relayed back to Earth stunning pictures of Jupiter and Saturn and their numerous satellites, and have visited the remotest parts of the solar system. It is frequently said that all these spectacular accomplishments have been the result of the development of science. This is true, but only partly so.

The obsession for transforming the world around us may be one of humankind's prime instincts that appeared long before the advent of science. Even at the dawn of history, people built roads, majestic temples, pyramids, and other structures that have struck the imagination of countless generations and even generated hypotheses about visitors from other worlds. The tales of extraterrestrials are myths of our epoch of the scientific and technological revolution, when people have ceased to believe in the possibilities of the purely inventive approach not based on scientific knowledge.

However, the pyramids, temples, clippers, diesel-engine vessels, locomotives, automobiles, and even planes are more the result of the spirit of invention than of systematic scientific research. The tree of science began to bear abundant fruit only in the twentieth century. The fruit, however, proved so significant as to dwarf all of humankind's previous accomplishments. During the twentieth century, science introduced two totally new technologies that have drastically changed the world in which we live: nuclear technology and electronics. All this happened during the lifetime of one generation. Now a third technology of the twentieth century—biotechnology—is developing before our very eyes.

As the appearance of transistors marked the beginning of modern electronics, the discovery of restriction endonucleases and the development of various methods of genetic engineering are contributing to the emergence of biotechnology. Like mushrooms after the rain, genetic engineering firms are springing up that utilize the most advanced technological principles to produce pharmaceutical preparations, vaccines, and other biologically active substances. What is the state of the art today and what will most likely occupy the huge army of scientists and engineers for the near future?

Let us start from the beginning. The passions that were raging around genetic engineering in the mid-1970s had been aroused not so much by its successes as by the apprehensions that it might have unpredictable negative consequences. The passions have raged unabated to this very day.

Is Genetic Engineering Dangerous?

It has been said that before the first test of an atomic bomb, the leaders of the Manhattan project asked the theoreticians these questions: "Will not such a totally new explosion, with no precedents in human history, cause a global catastrophe? Will not the atomic bomb serve as a detonator to set off a thermonuclear chain reaction that would envelop the whole of earth's atmosphere?" The theoreticians' first response was that most likely nothing terrible would happen. But what is the meaning of *most likely* in a situation like this? Ordinary acceptable allowances are totally irrelevant in answering such questions, for the fate of all humanity might well be at stake.

That was why, having pondered the question, the theoreticians decided that one of them should be charged with preparing a most accurate answer as to whether there was even the slightest possibility that such a catastrophe could occur. The choice fell on H. Brait, the most meticulous and levelheaded of the U.S. theoreticians, who would delve into every detail. Just imagine this colossal and dreadful responsibility placed on the shoulders of one person! After a most painstaking analysis of all the conceivable possibilities, Brait's answer was this: The possibility of an atomic blast causing a chain reaction in the atmosphere ought to be totally ruled out.

The birth of genetic engineering set the stage for a drama of similar intensity. In 1974, in the wake of the first experiments to obtain recombinant DNA molecules and prove their successful functioning in the cell, scientists asked themselves these questions: "What if gene reshuffling, which is quite impossible under natural conditions, results in the appearance of a DNA molecule with properties of unmitigated deadliness for man? What if the hypothetical molecule begins to replicate itself in an unrestrained fashion, infect huge masses of people, and eventually kill them all?"

A team of leading U.S. molecular geneticists, led by Paul Berg, made public, first in the main scientific journals, and then through the mass media, a sensational letter stating that they had interrupted their work on genetic engineering. They urged their colleagues around the world to follow suit, pending the convening of an extraordinary congress of specialists, which, they proposed, would decide whether the fears had any substance and work out measures for a maximum reduction of the risks involved in the activities pursued in genetic engineering.

Although the Asilomar Conference, held in 1975, had decided to ban activities in genetic engineering, the ban was lifted one year later. During that period, however, clear guidelines had been elaborated as to how researchers should proceed with genetic engineering activities involving various degrees of risk.

However, certain research activities, including those involving pathogenic microbes and oncogenic viruses, were totally banned. Special studies were also carried out to determine the degree of risk involved in various genetic engineering activities and to devise techniques that would practically eliminate all risks.

It is clear that in this case the problem cannot be resolved once and for all. Considering that potential genetic engineering risks are connected with an extremely complicated set of microbiological, ecological, and other factors, probably the only way to proceed here is through a gradual slackening of the restrictions and a careful checking of all consequences. Originally very strict, the rules for work with recombinant DNA were softened.

For the moment, all the conceivable verifications entailing enormous expenditures have failed to detect even the slightest impact of genetic engineering experiments on the microbiological environment around us. The fact is that recombinant DNAs lack all vitality when taken outside of the artificial conditions in which genetic engineers cultivate them.

There are grounds for believing that the situation is well under control and that if some nasty surprises do pop up, they will be detected before they become irreversible, and the potential threat will be eliminated. Whenever you use a lighter, a gas stove, or an electric iron (to say nothing of a nuclear reactor), some risk is involved. So it would be sheer stupidity to abandon research capable of solving many of humankind's acute problems just because of this consideration.

Now, years after the dramatic developments described, genetic engineers are proceeding full steam ahead in hundreds of laboratories around the world. Were the apprehensions justified? Had Berg's and his colleagues' appeal been nothing but a deft trick, as some cynics suggested, designed to attract to genetic engineering the attention of the public and of those who fund science?

On the one hand, the experience of the past years has demonstrated that, with appropriate precautions, genetic engineering activities do not involve substantial risk. On the other hand, however, humanity has found itself in these years faced with a new, appalling disease that has starkly demonstrated the terrifying insidiousness of viruses. At issue, as the reader may already have guessed, is the notorious AIDS (acquired immunodeficiency syndrome). This disease, of which humanity first became aware in the early 1980s, has inspired a terror akin to that felt by people in the Middle Ages due to plague or cholera.

It is true, AIDS reminds us how crafty and ruthless DNA can be. At the same time, there is no better example of the benefits that genetic engineering and biotechnology can provide than the recent dramatic success in fighting AIDS.

Battle of the Century

It is believed that this "plague of the twentieth century" originated in Central Africa. From there it was introduced into the Caribbean basin and was brought to the American continent via Haiti. In 1983–1984 French and American teams of virus experts were able to isolate the AIDS virus or HIV. The virus strikes at T lymphocytes (i.e., blood cells responsible for immunity). As a result, the patient loses immunity to any infection and may die of pneumonia or of some other disease. AIDS is mainly transmitted through blood and other bodily fluids, like serum hepatitis. Unlike hepatitis and all other diseases known to medical science, however, an AIDS victim faces the unavoidable prospect of death, for the immune system, which is a person's sole protector from viruses, becomes utterly disabled. Two factors have contributed to the generally belated detection of AIDS: The prolonged incubation period, during which the disease does not betray itself by any external manifestations, and the fact that death actually is due to other diseases. As a result, by the time the medical community braced itself for combat, about a million Americans were already carrying the virus.

Scientists around the world joined together to fight AIDS. Armed with the accomplishments of molecular biology and genetic engineering, they subjected HIV to wide-ranging investigations. The full nucleotide sequence of the viral RNA was determined. (The virus belongs to a class of viruses for which genetic material is provided by RNA, not DNA.) Special AIDS journals are published, and many periodic scientific editions regularly carry articles on the disease. An immunological test has been developed, permitting detection of HIV in the blood, identification of its carriers, and even the verification of donor blood.

Since the late 1980s, very significant resources have been directed to study HIV and to fight AIDS; however, researchers at first did not fully appreciate the enormity of the challenge. After all, since Edward Jenner invented vaccination 200 years ago, numerous viral diseases have been successfully defeated. The most dramatic success in recent history was the invention of the renowned Salk vaccine against polio. As a result, an infamous childhood scourge, which terrified many generations, was essentially eradicated.

In the case of AIDS, unprecedented efforts of researchers all over the world equipped with modern tools for creating vaccines seemed to yield results very quickly. Unfortunately, these efforts have repeatedly failed to produce good results yet. Jonas Salk himself spent the last eight years of his life trying to repeat for AIDS his spectacular success with the polio vaccine. He failed. As with the influenza virus, HIV has proven to be able to mutate too fast. Vaccination against one viral strain does not defend against another strain. Although the efforts continue and may lead eventually to success, the creation of a vaccine against AIDS remains as distant of a goal as it was ten years ago.

The failure in creating a vaccine emphasized the enormity of the challenge that researchers encountered in the case of AIDS. If a vaccine is unavailable, doctors are virtually helpless in fighting a viral disease. It is not surprising that by the early 1990s, AIDS patients virtually lost all hope for recovery. People became so terrified of the disease that even a positive HIV test was universally considered as a death sentence. In many cases, it immediately led to a change in the person's lifestyle, long before any symptoms of the disease emerged. The most publicized case of this sort happened with basketball superstar Magic Johnson, after he learned that he was HIV positive in 1991. Meanwhile, AIDS has continued to spread throughout the world, especially in Africa and Asia. By the end of 1996, the total number of HIV-infected individuals worldwide was estimated as 22.6 million; 3.1 million were infected in 1996 alone.

Despite disappointing news on the vaccine front, scientists have not given up. Only now, after failure to create a vaccine, have they realized that finding a cure for AIDS might not be accomplished by the traditional methods. A new approach to combat the viral infection had to be elaborated. It was clear that such an ambitious goal could be reached only as a result of a truly deep understanding of all stages of the HIV infection. A major hope was to find an Achilles' heel of the virus and hit it with a specially designed drug. However, never before had such an approach been used successfully to fight a viral infection.

Actually, a vulnerable point of the HIV virus was discovered very early in the course of its study. In 1987, a derivative of dideoxythymidine (AZT)—an inhibitor of reverse transcriptase—was the first drug to be used against AIDS (during the very early stages of the epidemic). Because synthesis of a DNA copy of viral RNA is a crucial step in the infection process, inhibition of reverse transcriptase was expected to prevent the infection. At the same time, reverse transcription does not play any role in human cells, so AZT was not expected to interfere with functioning of normal cells. However, although AZT gave promising results and slowed down the infection, it did not stop it altogether. The number of viral particles in the patient's body temporarily decreased and the number of T lymphocytes—target cells for HIV—temporarily increased, but after a while the deadly trend resumed. The main problem was again due to the vicious mutability of HIV: Among numerous mutants in the body, there were some more resistant to AZT than others. Therefore, in the presence of AZT, selection of resistant mutants of the virus occurred. Once this happened, AZT stopped being effective for that particular patient.

The situation was very similar to one doctors had encountered before with resistance of bacteria to antibiotics, which we already discussed in Chapter 5 in connection with plasmids. The radical difference was in the speed of acquiring the resistance. In the case of bacteria, the drug resistance emerged as a major problem only after several decades of massive use of the first antibiotic, penicillin. Millions and millions of lives had been saved by this time and new, more powerful antibiotics had been developed to replace penicillin. In contrast, resistance to the first anti-AIDS drug, AZT, appeared during the treatment of each patient.

Still, hope remained that a combination of drugs inhibiting reverse transcriptase (RT inhibitors) could be more efficient than AZT alone, or that drugs hitting another enzyme vital to viral development could be found. Researchers took both routes. Since 1991, five new RT inhibitors have received Food and Drug Aministration (FDA) approval. Doctors started to use various combinations of drugs or "cocktails," to try to stop the HIV multiplication before cocktail-resistant species evolved. Success remained quite limited.

A real breakthrough came after doctors began adding radically new drugs—protease inhibitors—to cocktails prepared for their patients. During their painstaking studies of HIV development in the body, researchers found an unusual pattern of "maturation" of HIV coat proteins, which form a capsule for the HIV genetic material, RNA. These proteins are first synthesized in the form of large polyamino acid molecules consisting of a linear array of several protein chains. Then a special virus-specific protease cuts the long chain, yielding separate proteins. If the protease is inhibited, the mature viral particle cannot form. The protease was subjected to intense scrutiny. Its structure was solved by X-ray analysis. After many unsuccessful attempts, various groups from major pharmaceutical companies developed several protease inhibitors that specifically inhibit the HIV protease. Three such protease inhibitors have been approved by the FDA in late 1995 and early 1996.

Now doctors have in their hands two types of drugs that inhibit two different but equally crucial enzymes necessary for HIV functioning. Any of these drugs alone, though demonstrating clear-cut effect, does not prevent HIV infection. However, when AIDS patients took cocktails consisting of different combinations of both types of drugs in high doses for a prolonged period of time, the number of viral particles in their bodies dropped dramatically and in many cases reached undetectable levels. During 1996, this miracle was repeated again and again with hundreds of patients in different clinics. Pioneer trials of the drug cocktails were performed in Aaron Diamond AIDS Research Center in New York with promising results. The center's director, David Ho, was selected as *Time* magazine's Man of the Year in 1996.

It appears that 1996 was a major landmark in human history. It will be remembered as the year when humans for the first time defeated a viral infection not by stimulating the means given to us by nature—our immune system—but instead by using the whole arsenal of powerful tools of modern molecular biology and biotechnology. In doing so, researchers figured out how to defeat a virus by striking the most sensitive links in its development cycle. It appears that if the attack on these links is massive and prolonged, multiplication of the virus is arrested before it has opportunity to create a defense against the attack.

Of course, we are only in the very beginning of a new era of combating AIDS. The treatment is painful and expensive. But for the first time since people confronted this terrible disease more than a decade ago, and after a series of disillusions and tragic losses of many lives, we clearly see the light at the end of the tunnel. All around the United States, patients are leaving the special AIDS hospices where they were supposed to spend their last days. Certainly, much remains to be done and many new obstacles will emerge. But at least being HIV positive no longer means a death sentence.

Among different genetic engineering techniques, without which the recent victory in combating AIDS would have been impossible, one occupies a special place. This technique, PCR (previously discussed in Chapter 3), made a real revolution in genetic engineering and biotechnology.

DNA Chain Reaction

It is very hard to admit that there is only one single reason for each of us to come into this world: to transmit our DNA to the next generation. There is absolutely no other purpose for us to be born. It is very uncomfortable to realize that our body is actually nothing more than a shell to carry DNA. There is no difference, with respect to this goal of existence, between a human and a bacterium, or a simple virus, or even a plasmid. From the biological viewpoint, people have been wandering in darkness trying to find the goal of their existence in cults, religions, music, poetry, and fine arts.

Although they all have the same goal, various species differ drastically with respect to the means they have at their disposal to reach this goal. Keeping in mind the simplicity of the goal, the diversity and the degree of sophistication that nature demonstrates seems truly amazing. However, if you think about it, you will realize that under conditions of fierce competition for limited resources, more primitive organisms should eventually lose to more sophisticated organisms, let alone different species. It still remains to be seen whether humans are sophisticated enough to avoid eventually following the fate of dinosaurs.

One can state that the above argument is correct only for rather sophisticated organisms, like animals, and that primitive organisms like bacteria, viruses, and plasmids compensate for their lack of sophistication by their ability to multiply with fantastic speed. This is partially true—but only if such multiplication does not harm humans. The most sophisticated creature on earth has developed and is continuing to perfect powerful tools capable of eradicating harmful germs. This growing arsenal of tools includes vaccines, antibiotics, and various drugs. In the previous section we have discussed our latest proud victory over the most crafty enemy that people have ever encountered, the HIV.

Although at present the human body looks like the best package for DNA multiplication, we still witness on our planet an impressive variety of DNA's packaging. The double-stranded nature of DNA is most suitable for fast multiplication. Indeed, if we separate the two DNA strands and synthesize complementary strands on both of them, we will have two daughter molecules identical to the mother DNA molecule. If we repeat the same trick with the two daughter molecules, we will have four granddaughter molecules identical to their grandmother. On the nth generation we will have 2^n molecules, all identical to the original single molecule. Such a process, which leads to exponential growth of the number of species, is called the chain reaction.

The term came from an important class of chemical reactions, which was discovered in the 1930s by the Russian chemist, Nikolay Semenov, and his disciples, Yuli Khariton, Yakov Zheldovich, and David Frank-Kamenetskii (author's father). Although such exponential growth had been well known in biology, chain reactions were totally new for chemistry. Semenov was awarded the Nobel Prize in chemistry for this discovery. The chain reaction proved to be crucial in chemical explosives. It was equally important in designing the nuclear reactor and the atomic bomb. Because of the publicity that surrounded the atomic bomb, the term "chain reaction" came into general use.

Life, therefore, may be considered a DNA chain reaction. This chain reaction proceeds in a controlled manner (like the nuclear chain reaction does within nuclear reactors), when the birthrate is approximately equilibrated by the death rate and the quantity of creatures is more or less constant with time. However, sometimes the DNA chain reaction resembles an explosion, for example, during epidemics of infectious diseases.

The analogy between chain reaction and the multiplication of living things has been obvious since the discovery of chain reactions in chemistry. Chain-reaction-type DNA multiplication became clear immediately after the discovery of the double helix. What is amazing is that before the mid-1980s no one tried to perform the DNA chain reaction in a test tube, although everything was in place. DNA melting (i.e., the separation of DNA strands by heating) was a well-studied process by this time. Synthesis of DNA primers had become routine business. DNA polymerase I, which extends primers, was quite available and widely used (see Chapter 6). Researchers in academia just did not see a reason why they needed to multiply DNA. In one of his papers, Khorana mentioned that the DNA chain reaction can be performed in the test tube via periodic heating and cooling of the sample in the presence of primers and four dNTPs, using DNA polymerase I. So what? He did not want to waste his time showing that the idea would work. Of course, it would!

Most likely, Kary Mullis did not read Khorana's paper. Mullis worked at Cetus Corporation, one of many biotechnology companies that mushroomed in the late 1970s and early 1980s. Being in industry, he clearly understood that the possibility of multiplying DNA in a test tube could produce a revolution in biotechnology. So enthusiastic was he about his invention of PCR, that he convinced his colleagues at Cetus to begin experimentation. The scheme of their first experiments is shown in Figure 31.

First, a target sequence was chosen on DNA. The sequence had to be known, or at least its two termini. Then two DNA primers were synthesized. One was complementary to the left end of the target sequence of the bottom DNA strand; the second one was complementary to the right end of the target sequence of the top DNA strand. These two primers were added to the sample in huge excess as compared with the number of DNA molecules. (PCR actually may be conducted starting with a *single* DNA molecule.) All four dNTPs were also added in sufficient quantity. The sample was heated to a temperature that guaranteed DNA melting (i.e., the separation of the complementary strands). Then the sample was cooled. During the cooling stage, the synthetic primers found complementary sites on separated DNA strands, whereas the two long DNA strands were unable to find each other because they were present in minute concentration.

Thus, as a result of the first cooling, two substrates were prepared for the primer extension reaction (see Chapter 6). Therefore, when DNA polymerase I was added to the mixture, it extended the two primers in opposite directions. As a result, two daughter DNA molecules appeared. They were both only partially double stranded and carried long single-stranded tails. However, the target sequences in both molecules were fully double stranded. Further heating/cooling/polymerase-adding cycles led to the

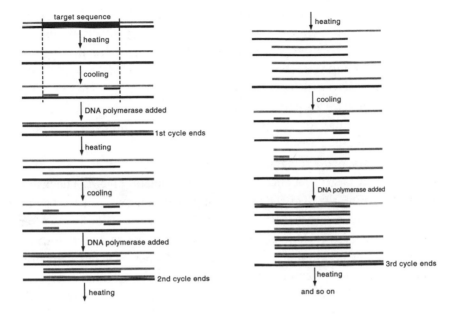

Figure 31. Three cycles of the polymerase chain reaction (PCR).

synthesis of more and more molecules, all in which the target sequence was double stranded.

As Figure 31 shows, only at the 3rd cycle do first molecules appear, which exclusively consist of the target sequence. They are the molecules that exponentially multiply during the subsequent cycles. There are 8 such molecules at the 4th cycle, 32,738 at the 15th cycle, and as many as a billion at the 30th cycle. In their first experiments, the Cetus team could not afford that many cycles. A major problem was the inactivation of the added enzyme due to heating, so that new portions of the enzyme had to be added again and again at each cycle.

A major breakthrough, which made the PCR such a brilliant technique, consisted of replacing DNA polymerase I with the so-called *Taq* DNA polymerase. Extracted from thermophilic bacteria, *Taq* polymerase is not inactivated by being heated up to 94°C, a sufficiently high temperature to melt DNA. *Taq* polymerase loves hot conditions, so the primer extension reaction is conducted at 72°C.

Implementation of *Taq* polymerase made it possible to perform the PCR in a fully automated manner in simple robots called thermocyclers or

PCR machines. A mixture of target DNA, primers, dNTPs, and *Taq* polymerase is prepared and then placed within an ordinary test tube in the thermocycler, which has been programmed to perform three cycles: heating up to 94°C (for one minute), cooling down to 60°C (again for one minute; for the primers to anneal to their sites on DNA), and then heating up to 72°C (again for about a minute) to perform the primer extension reaction under the optimal temperature for *Taq* polymerase. Then the full cycle is repeated again, and again, and again.

Among the many remarkable features of the PCR, one is that you need not purify your target sequence from any contaminating DNA. The primers strictly determine the sequence to be amplified and you can start with a single copy of your target sequence in a huge excess of other DNA. After a sufficient number of cycles, you will have as much target DNA as you wish.

For his invention, Mullis was awarded the Nobel Prize in Chemistry in 1993. The invention is considered to be one of the very few major breakthroughs in the history of DNA technology, which, as we know, is full of remarkable discoveries.

The reason for such astounding success of an essentially very simple, if not trivial, idea is twofold: First and foremost, PCR works remarkably well, much better than Mullis could have dreamed when the idea occurred to him. Second, such a good method of DNA multiplication proved to be indispensable in numerous applications. If the biological essence of life is the DNA chain reaction, artificial life is now flourishing in thousands of PCR machines in laboratories all around the globe.

Genetic Engineering Pharmacology

Pharmaceutical companies were the first to take a serious interest in genetic engineering. As they saw it, the prospect of producing practically any protein in large quantities and at a comparatively low price meant undreamed-of opportunities. Indeed, in addition to being the cell's chief "working molecules," proteins also have a key role to play in regulating processes at work in the organism as a whole. Almost all hormones are small protein molecules, with the number of amino acid residues ranging from one to several dozen.

Previously, the production of hormones was frequently a delicate operation. It is best if, as in the case of insulin, an animal protein (of large cattle or pigs) can serve as a substitute for a human hormone. But many times this is not possible. Pharmaceutical companies thus found the prospects of genetic engineering irresistible. On their orders, genetic engineers obtained, in a brief period, strains of bacteria that produced different human hormones.

One example is that of growth hormone. Some children's bodies do not produce growth hormone because of a genetic defect. If left untreated, these children turn into dwarfs. They are clearly in dire need of this hormone, which had previously been possible to obtain only from human corpses. Genetic engineering, however, has come up with a method for the large-scale production of this hormone.

Another example is that of insulin. Insulin is, of course, a hormone that has a wide application, because it is needed by those suffering from sugar diabetes, a very widespread disease. Although most diabetes patients manage to do nicely with animal protein, some are allergic to it and need human insulin. Again, genetic engineering made it possible to produce human insulin.

The most exciting challenge, however, was the prospect of obtaining human interferon. Although the subject of lively debate for more than two decades, interferon has remained an enigma in many ways. The only thing that had been firmly established about interferon is that it is a protein that is highly effective against the most diverse kinds of viruses.

One can say that interferon's action against viruses is the same as that of antibiotics against bacteria, with one important difference. An antibiotic effectively suppresses bacterial proliferation in any organism if the bacteria in question lack the requisite genes of resistance. For its part, interferon is very finicky in that a virus infection in the human organism can be suppressed only by human interferon (or by that of our close relative, the ape). Although the control of viruses (that are totally insensitive to antibiotics and would generally succumb only to vaccines) remained the main problem, another was that it had been impossible for quite a long time to obtain pure interferon in sufficient quantities and at a relatively cheap price. In fact, interferon's amino acid sequence had defied all attempts at reading it. Then in a matter of one year, genetic engineering drastically changed the situation.

In the case of interferon, researchers were able to apply two techniques to induce the cell to produce an alien protein, as discussed in Chapter 5. In the first, genetic engineers proceeded as follows: They isolated interferon mDNA from human blood cells in which the production of interferon had been stimulated by a virus infection. Then, using reverse transcriptase, they synthesized on the molecule an interferon gene and, incorporating it into a plasmid, created the first bacterial strain to produce artificial interferon at very high rates. The full amino acid sequence of interferon was finally established.

The second technique was a purely chemical one. The amino acid sequence permitted the nucleotide sequence of the interferon gene to be deduced with the aid of the genetic code. Then the artificial gene was synthesized by chemical methods. The gene was also incorporated into a plasmid, thus creating another interferon-producing strain.

Artificial interferon proved to be a powerful antivirus preparation, as was shown in the following experiments. Six monkeys were selected and divided into two groups of three. The virus of encephalomyocarditis was introduced into all the monkeys; lacking immunity to this virus, they were all doomed to die. In fact, the three monkeys in the control group died several days following the infection. The monkeys in the second group, however, had received artificial interferon four hours before the virus was introduced, and several times following infection. They all survived.

Artificial interferon made it possible to embark on a broad course of biological and clinical testing of the preparation. These tests have recently led to approval of interferon as a medication in cases of certain viral diseases, such as hepatitis and sexually transmitted diseases induced by the papilloma virus. Without genetic engineering, interferon would have remained until now, and for a long time to come, an enigmatic, albeit very promising, protein.

The production of vaccines has become another broad area of application for genetic engineering in medicine and agriculture. Vaccination is the most effective means for the prevention of viral epidemics. Ordinarily, use is made of dead viruses whose DNA (or RNA) has been destroyed through a particular procedure, but whose proteins have been preserved. When the dead viruses are implanted in the organism, the organism begins producing antibodies against the proteins, so that if subsequent "living" viruses penetrate into the organism, they are recognized by these antibodies and rendered harmless by the immune system.

Vaccination heralded the total elimination of many diseases that used to kill millions of people. There are still viruses, however, that have so far remained undefeated. For human beings, the worst are the viruses that cause AIDS and influenza; for animals the worst is the virus that causes foot-and-mouth disease. Attempts to control these viruses with vaccines have met with only limited success.

One reason for this has been the high degree of mutability shown by these viruses. They mutate frequently, with their proteins affecting substitutions of individual amino acids and the "old" antibodies losing their ability to recognize these proteins. The consequence is that vaccinations must be repeated over and over again. Frequent vaccinations on a vast scale have one major drawback—namely, that it is difficult to ensure the full inertness of the vaccine (i.e., to make sure that absolutely all the viral particles in the preparation injected have been killed). If all the viral particles are not killed, the vaccine, rather than acting as a savior, may spawn an epidemic and turn into a killer. Paradoxical though it may be, according to media reports, most of the foot-and-mouth disease epidemics at present are caused by less-than-perfect vaccines.

In principle, genetic engineering is capable of making absolutely harmless vaccines. One only needs to make bacteria produce one protein (or several) of the viral coat, and then use that protein for vaccination. In this case, the vaccine carries no infecting agents (DNA or RNA), and so it can set immunity to work without causing the disease. A vaccine of this totally new type was obtained and tested. The experiments were conducted with one of the coat proteins of the foot-and-mouth disease virus. The results were not bad, although it turned out that the immunization achieved with this vaccine showed an effectiveness that was only about one-thousandth of that achieved when the killed virus was used to make the vaccine.

Many epidemiologists believe, however, that such radically new vaccines will not find broad application. Their skepticism is due to the fact that such virus-induced diseases as hepatitis and AIDS are more widespread in developing countries where the level of medical care is as yet incapable of absorbing these too-novel and complex vaccination techniques. They also cite the fact that the greatest single success in curbing a viral disease on a world scale has been the eradication of smallpox through the use of live vaccine.

The beginning of the vaccination success story, in which epidemiologists take proper pride, dates back to the time when smallpox was still taking a heavy toll on human lives in Europe. In 1798, an English physician, Edward Jenner, noticed that milkmaids who had been ill with a light form of smallpox they had gotten from cows no longer fell ill with the disease. He began to infect healthy people deliberately with cow smallpox and thus protect them from the full-blown disease. This was the beginning of vaccination. (The very word *vaccine* in Latin means "of the cow.")

Much later, when, thanks to Jenner's invention, smallpox was practically eradicated in Europe, it was found that both types of smallpox were caused by viruses. Although different, the viruses were related. Some proteins located on the surface of the cow virus, known as the smallpox vaccine virus, are identical to surface proteins of the smallpox virus. Thus, the immune system, placed on the alert after vaccination with the smallpox vaccine virus, ensured excellent protection against the smallpox virus as well.

The smallpox vaccine virus proved to be a perfect find for epidemiologists. It is practically innocuous to humans, is highly effective at immunization, and multiplies easily in cows infected with it. All of this enabled the World Health Organization (WHO) to conduct a broad and prolonged smallpox-control campaign that was crowned with brilliant success. In 1977, the WHO announced that the disease, which not so long before had killed people by the millions, was completely eradicated. In 1988, WHO embarked on another, even more ambitious project by establishing the goal to eradicate the polio virus by the year 2000. Most likely, this objective will be reached. Both of these are examples of successful implementation of traditional, pregenetic-engineering vaccines.

B. Moss and his research assistants from the National Institutes of Health decided to modify the smallpox vaccine virus by using genetic engineering techniques, so that it would protect people from both smallpox and hepatitis. They built a surface-protein gene of the hepatitis virus into the DNA of the smallpox vaccine virus, having equipped it with an effective promoter. Experiments conducted on rabbits demonstrated that hepatitis protein was produced after vaccination with the virus and that, in response to the production of this protein, many antibodies against the hepatitis virus began to appear in the blood.

Based on the smallpox vaccine virus, Moss is going to make a live vaccine for a whole range of virus-induced diseases by building into the virus's DNA the genes of the appropriate surface proteins. If the testing proves successful, the WHO will again be able to launch a program similiar to the one that proved so successful in the case of smallpox. There will be no need to retrain the medical personnel—they will be able to work with the familiar smallpox vaccine virus. If Moss's idea is realized, then it may well become possible to kill several birds with one stone—namely, to put an end to several viral-induced diseases simultaneously.

The Coming of the Golden Age

Genetic engineers have embarked on a fresh crusade against viral diseases. There is good reason to expect that this resolute offensive will bring about a breakthrough in medicine and veterinary science similar to the one caused by the discovery of antibiotics. Indeed, biotechnology's impact on human life will not be confined to medicine alone. So far, however, it is very difficult to predict the impact of this breakthrough in other areas.

Although in the medical and veterinary fields it looks as if many things are progressing swimmingly, elsewhere things are still basically confined to nebulous promises of a golden age. Nevertheless, one task is already taking a sufficiently clear-cut shape—namely, the industrial production of protein as animal fodder. Conventional animal fodder, like hay or the green mass of corn, lacks protein, especially some of the amino acids. Compensation for this deficit means a drastic enhancement of the efficacy of conventional fodder. This has been known for quite a long time, and for many years the industry has been using microbiological methods to produce amino acids that are then added to fodder. Genetic engineering makes it possible to devise strains with an unprecedented productivity, which will undoubtedly help resolve the task of producing fodder with an optimum balance of protein components.

However, I personally perceive nothing new in all this. With varying success, the problem was being tackled even without genetic engineering. Whether the total transition to the industrial production of animal feeds using genetic engineering techniques will prove profitable is for the future to decide. If biotechnology proves this endeavor to be viable, it will amount to a veritable revolution. I, for one, envision a golden age that looks something like this.

Somewhere in the deserts stand solar electric power stations churning out electricity that, together with minerals from the bowels of the deserts, is supplied to huge biotechnological factories. Using bacteria or yeast, these produce artificial food and optimally balanced fodder, which is shipped in convenient-to-use packages to factories raising poultry, pigs, and cows.

There, in incubators like the ones used today to hatch chickens, they raise animals, perhaps including totally new ones created with the help of genetic engineering.

Conventional farming, which raises crops such as wheat, has been preserved within certain limits. But the demand for these quite-costly agricultural products has declined to such an extent that they are being produced only in certain climatic zones with perfect irrigation and other conditions. Vast tracts of land used for farming in pre-biotechnological times have been made available to people who have left the crammed cities for the wide-open spaces, with forests, rivers, and lakes, and use electromobiles to get to work, shop nearby, or to see friends.

Sooner or later, a new technology always changes everyday life, but it is very difficult to predict how this will come about. Electronics, for instance, has already brought about a radical change in the once usual modes of obtaining and processing information. The pace of these changes has rapidly increased recently because of the Internet. With electronics' passage to a qualitatively new stage—that of miniaturization—it is penetrating all spheres of life. Will this happen with biotechnology, too? The answer is an absolute and unequivocal yes.

CHAPTER 11

The Controversy Around the Double Helix

ooooo

Are Watson and Crick Right?

In our time, the word DNA has become as common as oil or steel. A great boom has already set in around DNA, with scores of laboratories and biotechnology companies engaged in hectic activity to produce "recombinant DNA" and large armies of specialists manipulating genes and looking for practical applications of the results of their manipulations. In the courts, gels showing bands have become as common as fingerprints. It is amazing to think that all this started on April 25, 1953, with a tiny article in the journal *Nature*, signed by two then largely obscure names: James Watson and Francis Crick. In the article the authors expounded their view of the structure of the molecule of the deoxyribonucleic acid. They reported that the molecule consists of two antiparallel polynucleotide chains twisted into a right-handed double helix, that the inside of the double helix contains nitrogenous bases forming the core of a sort of cable, and that the cable's coat consists of negatively charged phosphate groups. Nitrogenous bases of the two strands form pairs in accordance with the complementarity principle, with adenine (A) always matched by thymine (T), and guanine (G) matched by cytosine (C), as diagrammed in Figure 32. The base pairs are strictly perpendicular to the double-helix axis, like the rungs in a helically twisted rope ladder.

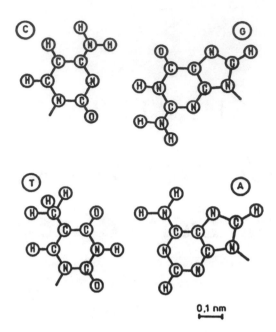

Figure 32. This is what the Watson–Crick base pairs look like. The scale in respect to distances between nuclei is correctly shown but atoms' sizes are greatly reduced. Indeed, the bases in the pairs cannot be brought more closely together, since then the hydrogen atoms of one base would overlap with oxygen and nitrogen atoms of another. One remarkable feature of the structure of the Watson–Crick pairs is that both have almost precisely similar sizes (distances between nitrogen atoms bound with sugar). It is because of this, that any sequences of base pairs can be "inscribed" within the double-helix structure.

This structure, which (according to a universal conviction) DNA assumes under physiological conditions, was named the B form. The DNA structure undergoes a substantial change only if the molecule is placed in a totally unusual environment, say, in a very concentrated (almost 80%) alcohol solution. On the other hand, as numerous findings indicate, the DNA structure remains practically unchanged when exposed to a broad range of external conditions.

Strange though it may seem, "cast-iron" proof that DNA is in fact a double helix had long been lacking. In fact, the experimental findings relied upon by Watson and Crick and their followers can be interpreted in more than one way. There is always the possibility that, within the limits of experimental accuracy, the same results may be found for some totally different DNA structure.

In the late 1970s, for instance, there was quite a hubbub over the model constructed by researchers from New Zealand and India, according to which the two DNA strands, rather than being twisted around each other, run parallel to one another. This model was known as the side-by-side (SBS) form.

Originally, it was asserted that the X-ray diffraction pattern of the SBS form was the same as that of the B form. When it became clear that this was not so, however, the researchers began to assert that in fibers, in which DNA was studied using the X-ray analysis, DNA can have the B form; whereas in solution, and even more so in cells, it undoubtedly assumes the SBS form. The absence of topological problems (i.e., it is not necessary to unwind the helix in the course of DNA replication) was regarded as a strong point of this model.

That the SBS form does not stand on its own as a DNA model under ordinary conditions has been demonstrated by many techniques. Nevertheless, the model had given rise to a debate that eventually proved very useful. It prompted researchers to adopt an even more painstaking approach to their certainty that the Watson–Crick model holds true in all its basic components, not just that DNA consists of two strands whose sequences are mutually complementary.

The most convincing evidence to support the Watson–Crick model came from experiments done with circular DNA. These experiments, first conducted by Harvard's James Wang, showed that DNA was really a helix, and made it possible to estimate with a high precision the pitch of the helix. However, Wang's experiments, although absolutely convincing, required a complicated analysis. Therefore, instead of explaining Wang's experiments, the more obvious experiments of D. Shore and R. Baldwin at Stanford University will be described.

Shore and Baldwin investigated the probability of a DNA molecule closing into a ring based on its length. To clarify the problem, they took molecules with cohesive ends (such molecules were already discussed in Chapter 9) and added DNA ligase. They judged the probability of closing by the appearance of closed circular molecules. At first, Shore and Baldwin confined themselves to natural molecules; then they employed genetic engineering techniques that enabled them to study very short chains containing a mere 200 pairs. At first, the resulting picture was quite gratifying to the researchers: It squared with theory and common sense. In the case of very long molecules, containing many Kuhn statistical segments, the closing probability decreased with the *growth* of the chain length. Conversely, in the case of short molecules, the closing probability decreased with the *reduction* of the chain length. This is quite understandable, since the long molecules look like the itinerary of a man who lost his way in the forest, while the short ones are like a policeman's rubber baton—the shorter the baton, the harder it is to bend it into a ring.

One circumstance confused the researchers. As the length decreased, the results grew widely scattered, although the experiments with short DNA were staged as painstakingly as those with the long ones. What was the matter? To understand this untoward situation, Shore and Baldwin prepared a set of specimens, based on genetic engineering techniques, which contained molecules with 237, 238, and so on, up to 255 nucleotide pairs. When they measured for each preparation the probability of formation of circular chains and entered the data into a chart, they obtained a sinusoid section with a period of 10 pairs. The reason for the scattering of points became clear; it turned out to have been caused not by accidental discharges, but by regular oscillations related to DNA's helical structure.

To understand the results of these very important experiments, let us imagine that we are dealing with a circular DNA with one of its strands nicked and "left to its own devices" in solution. What will things look like at the nick point? It may well be that the nicked strand is ready for "docking," and you only need to form a link as shown in Figure 33a. At the other extreme, the ends of the nicked strand may be incapable of making contact, as diagrammed in Figure 33b.

Now, let us add the DNA ligase, which cures the nick. The enzyme can do its trick only if the severed ends are in an abuttable position. Then what? Will it stitch together only such molecules? The answer is no. After all, a DNA molecule is a microscopic object. One hallmark of micro-objects, distinguishing them from macro-objects, to which we have become quite accustomed to in everyday life, is that micro-objects undergo significant changes in their size and shape simply as a result of thermal motion. In our macro-scaled world these changes are imperceptible; we simply do not see them.

We have already said that thermal motion bends linear DNA, preventing it from assuming the shape of a straight line. The same thermal motion

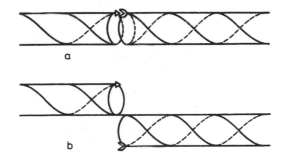

Figure 33. Successful (a) and unsuccessful (b) docking of a circular DNA molecule with a single-strand break.

prevents circular DNA from assuming the most energetically advantageous form of a ring. The molecule assumes in space a quaint, constantly changing shape. Additionally, thermal motion causes the angle of the twist between the two neighboring base pairs in a double helix to change constantly. Thus, in the Shore–Baldwin experiments, the ligase sealed the gap not only in nicked strands such as those depicted in Figure 33a, but also those depicted in Figure 33b and all intermediate cases. However, in the case of an abuttal favorable for "stitching" by a ligase (Figure 33a), the probability of closing was maximal, whereas in the case of an unfavorable abuttal (Figure 33b), it was minimal. The sinusoidal variation of the closing probability therefore reflected the rotation of one end of the nicked molecule with respect to the other, with the increasing DNA length due to the helical nature of DNA. The period of the sinusoidal curve corresponded to the pitch of the double helix. Through these experiments, Shore and Baldwin were able to provide a visual and obvious demonstration of DNA's helical structure and to assess the period of the helix. Subsequently, D. Horowitz and J. Wang determined the period of the helix with a very high degree of accuracy, based on data for short rings. It turned out to be 10.54.

One remarkable feature of the results obtained by both groups in these experiments, is that they were attained with isolated molecules in a solution. Indeed, from the time of the classical experiments of Wilkins and Franklin, and before the development of the DNA X-ray crystallography (see the next section), all information about DNA's detailed structure was obtained on the basis of X-ray data received with fibers in which molecules strongly interacted with one another.

Thus, the fact that DNA takes the B form in a solution has been proven beyond any reasonable doubt. This, however, is true only with respect to nonsupercoiled DNA. In a supercoiled state, the structure of the molecule's principal part does not undergo any appreciable change, but some sections with specific sequences can change their structure drastically. Let us recall the palindromes and cruciform structures whose existence has been proven experimentally. What, if any, changes can there be in the DNA structure? Do these changes play a biological role?

The Z Form

We have already mentioned that in modeling DNA's structure, Watson and Crick and their followers relied on data regarding the X-ray scattering patterns of DNA in fibers. This is precisely the structure with fibers, but not crystals, since natural DNA molecules extracted from the cell do not crystallize (they are too long to form a crystal).

With a partial drying of the solution, a certain "compacting" of the molecules occurs: They become stacked haphazardly like logs in a creek during timber rafting; not in two dimensions, as on the surface of water, but in three. The spaces, as in a creek, are filled with water. The X-ray scattering pattern obtained from such a partially ordered positioning of the molecules provides us with rich information. Unfortunately, it is not sufficient to reconstruct the molecule structure. This fact was the main point of contention in the argument over whether Watson and Crick had "guessed" right as to DNA's structure in fibers.

Wang's experiments contributed to the development of a situation that was, to a certain extent, paradoxical. It was now clear that in solution, isolated DNA molecules have a structure that basically corresponds to the Watson–Crick model. What about the molecules' structure in fibers where the molecules interact with each other? Does this structure change? Specialists engaged in DNA modeling and in calculating the X-ray scattering pattern depending on the model, confidently asserted that only the B form of DNA could create the pattern observed. There could, however, be lingering doubts as to whether the researchers might not have overlooked something.

The problem could have been resolved if it had been possible to obtain crystals of DNA, study the X-ray scattering pattern from these crystals, and then solve the structure. Precisely this technique is employed in determining the spatial structure of ordinary chemical compounds of any complexity, and also of proteins. As applied to DNA, however, the technique has failed repeatedly.

Therefore, it became clear that there was no chance of obtaining crystals in the case of either long molecules or short molecules having differing lengths and sequences. There was, of course, the lingering hope that if one took short molecules of the same length (say a dozen base pairs) and sequence, and, one could somehow crystallize them. Obtaining crystals, however, is a painstakingly complicated task. You have to vary the stock solution from which you crystallize molecules; therefore, you need very large quantities of substance. Where can you find numerous DNA bits of a strictly determined length? Preparations of this kind became available only toward the close of the 1970s, due to the astonishing breakthroughs made in the chemical synthesis of DNA.

Accomplishments of chemists in this field have been truly spectacular. Thirty years ago, Khorana's synthesis of trinucleotides of different sequences caused quite a sensation, eventually earning him a Nobel Prize. (As the reader will probably remember from Chapter 2, these trinucleotides permitted the complete deciphering of the genetic code.) Today you can buy a box the size of a typewriter, equipped with A, T, G, and C push buttons. You push the buttons in the order in which you wish to have your sequence (not very long, of course); fill the box with the necessary ingredients, also available on the market now; and take time out to have lunch, go to the library, or attend a seminar. When you return several hours later you will discover the output of your box—several milligrams of the needed preparation (i.e., a synthetic piece of single-stranded DNA with the sequence that you "ordered").

This miracle of chemical and engineering thought resolves all the problems connected with the artificial synthesis of the gene. Using sections of, say, twenty nucleotides each, you can stitch together a gene of any length with the help of DNA ligase. This also resolves the problem of obtaining large quantities of short DNA bits for their crystallization. Incidentally, such equipment appeared in the early 1980s, but as early as the end of the 1970s, researchers in some laboratories that were engaged in gene synthesis could rapidly synthesize DNA, albeit manually.

Alexander Rich and his co-workers (MIT) were the first to obtain good crystals of synthetic DNA. The crystals obtained were of a hexanucleotide:

C G C G C G

G C G C G C

You may well imagine the excitement of Rich and his co-workers when, having performed all the necessary and extremely labor-consuming operations, they finally obtained the structure of their hexanucleotide. The structure had nothing in common with that of the Watson–Crick model!

Naturally, it had normal GC pairs and was even a double helix, but with twelve rather than ten base pairs per helical turn. The principal difference, however, was that rather than being right-handed like the B form, the helix was twisted left-handedly! Actually, almost a decade before the events we are speaking of, two researchers from Germany, Friz Pohl and Thomas Jovin, claimed, on the basis of their spectropolarimetric data, that a double helix consisting of alternating GC and CG pairs undergoes a transition into a left-handed helix in a highly salted solution. However, because they relied on indirect evidence, very few were convinced. Now, however, with X-ray crystallographic data, nobody can doubt that the left-handed double helix really exists.

There were other important differences between this novel DNA form, named the Z form, and the B form. One difference (from whence its name had come),was that unlike the B form, whose sugar–phosphate chains produced two smooth helical lines, the Z form had zigzag lines (Figure 34).

Does this mean that the Watson–Crick model was proven wrong in the final analysis? For the fact is that the very first DNA structure, as established by the absolutely reliable methods of X-ray crystallography, turned out to be quite different from the B form.

Despite all the uproar caused by the discovery of Rich and his co-workers, its ramifications were far from being that radical. Experiments with circular DNA have provided unequivocal evidence that the DNA helix in a solution is twisted right-handedly, with ten base pairs per helical turn, which corresponds to the B rather than the Z form. Could it be then, that intermolecular interactions are responsible for such a drastic change in the double-helix structure in crystals? This does not seem to be the case either.

In actuality, it appears that one factor is primarily responsible for this hexanucleotide's assuming of the Z form—the strict alternation of the G and C residues.

Richard Dickerson and his colleagues from UCLA found that DNA assumes the B form if the sequence is different than the alternating G and C

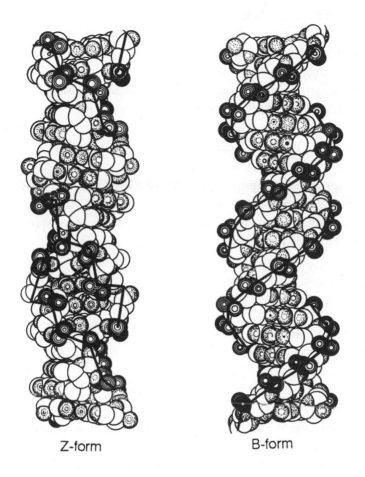

Z-form   B-form

Figure 34. Diagrams of space-fill models of Z and B forms of DNA. The black lines show the helical path of the sugar–phosphate backbone.

pattern. They checked the structure in a DNA crystal with a different sequence, a dodecamer:

C G C G A A T T C G C G

G C G C T T A A G C G C

and found the B form. Many researchers around the world began to study DNA's transition into the Z form in solution. It turned out that under ordinary conditions, at least in linear DNA, the Z form does not appear. But what about the case of a supercoiled DNA?

Supercoiling must, of course, make the Z form more advantageous, since the change in the helix's sense from positive to negative in a DNA section tends to remove tension in the rest of the negatively supercoiled molecule. It would thus be quite natural to assume that in supercoiled DNA, sections with alternating G and C sequences would transform into the Z form. But does this really occur?

The answer hinges on the energy required for a B-to-Z transition of the DNA section with the sequence . . . CGCGCGCGCG . . . . The fact is that (besides assuming the regular B form) assuming the Z form must be more advantageous than assuming a cruciform to enable the Z form to exist. As it happens, the sequence

C G C G C G

G C G C G C

is a palindrome pure and simple. Thus, the question of whether or not DNA sections with a suitable sequence assume the Z form is not that simple. Only experimentation could provide the answer.

As was the case with cruciforms, the technique of two-dimensional gel electrophoresis proved to be the most effective method of finding out whether or not the Z form is produced in the negatively supercoiled DNA. However, quite long sections . . . CGCGCGCG . . . are not encountered in common plasmids. Thus, it proved necessary to construct specialized plasmids carrying artificial . . . CGCGCGCG . . . inserts of different lengths.

Wang was the first to apply the technique of two-dimensional gel electrophoresis to the study of the Z form. Investigating plasmids with long . . . CGCGCGCG . . . inserts, he observed patterns of the type shown in Figure 27 under conditions in which no discontinuity on the electrophoregram was observed for the control plasmid with no insert. This signified that the observed structural transition occurred in the . . . CGCGCGCG . . . insert. But what appeared in the process: a cruciform or a Z form? The answer could be provided by the mobility drop value (i.e., by how many topoisomers the mobility drop occurred during the transition). If the outcome of the transition was a cruciform, one would expect a leap of $m/10.5$ topoisomers, where m is the number of base pairs in the palindrome (i.e., in the . . . CGCGCGCG . . . sequence). If the outcome was the Z-form structure, however, one would expect a leap of $m(1/10.5 + 1/12.5)$ topoisomers (12.5 is the number of pairs per turn of the left-handed DNA helix in the Z form).

In the experiment performed by Wang, the drop value was shown to be related to the formation of the Z form and not to a cruciform. It was thus demonstrated that . . . CGCGCGCG . . . sequences can flip into the Z form in conditions that are close to physiological. This made it possible to hope that the Z form may arise in DNA inside the cell and play some biological role.

The appearance of the Z form in CG sequences with negative DNA supercoiling can be obtained by techniques other than two-dimensional gel electrophoresis. This alternate method, broadly utilized by Rich and others, consists of the use of antibodies to the Z form. These antibodies were obtained by immunizing animals using the chemically modified CG polymer that is present in the Z form under all conditions. Such antibodies do not bind with DNA or the CG polymer in the B form, but strongly associate with the CG polymer in the Z form. It was demonstrated that artificial plasmids that carry CG inserts begin to bind antibodies to the Z form, when their negative supercoiling becomes sufficiently high.

Analysis of the spatial structure of Z-form DNA led to the conclusion that a regular alternation of purine and pyrimidine nucleotides in either of the complementary strands is important. In the absence of such an alternation, the Z form becomes highly disadvantageous when compared with the B form. Thus, one would expect that, along with . . . CGCGCG . . . sequences, negative supercoiling would be conducive to the Z-form structure in two other simple purine–pyrimidine sequences:

$$. . . G\ T\ G\ T\ G\ T\ G\ T\ G\ T . . .$$

$$. . . C\ A\ C\ A\ C\ A\ C\ A\ C\ A . . .$$

and

$$. . . A\ T\ A\ T\ A\ T\ A\ T\ A\ T . . .$$

$$. . . T\ A\ T\ A\ T\ A\ T\ A\ T\ A . . .$$

It was especially important to clarify this issue, because long sections with such sequences turn up with comparative frequency in eukaryotic DNA. Special plasmids were constructed, carrying such inserts, and experiments were conducted with two-dimensional gel electrophoresis. The experiments predictably proved that, in a negatively supercoiled DNA, GT inserts transition into the Z form (formation of cruciforms in such sequences is apparently not possible), while AT inserts transition into cruciforms.

The discovery of the Z form caused quite a stir among molecular biologists. In addition to the formation of cruciforms in superhelical DNA, the discovery demonstrated that although DNA as a whole is, without doubt, structured in the B form, its individual segments may have drastically different structures. Thus, a search began to find these and other structures in DNA, and clarify their possible biological role.

The H Form

The most popular technique for clarifying the possible biological role of alternative (different from the B form) structures proved to be the enzyme method, because it is simple and permits localization of the sites on DNA that are attacked by the single-strand-specific endonuclease. Lilley demonstrated that cruciforms are attacked in the palindrome center, and Wells also showed that, during the appearance of the Z form, the attacks are launched between it and the B form.

After building into plasmids newer DNA segments from the most diverse organisms, genetic engineers began to test them for sensitivity to the single-strand-specific endonuclease in the hope of discovering cruciforms or the Z form. Indeed, certain DNA segments of higher organisms proved to be very sensitive to the enzyme. These were called hypersensitive sites. When it became possible to pinpoint these sites, it was discovered that they were always located in important regulatory sections of genomes. But when the hypersensitive sites had been sequenced, the researchers found to their great confusion that the sections were neither palindromes nor alternating purine–pyrimidine segments. As a rule, hypersensitivity to the single-strand-specific endonuclease was exhibited by sequences that contained only purines in one strand and only pyrimidines in the other—that is, homopurine–homopyrimidine sequences of the $(G)_n \cdot (C)_n$ or $(GA)_n \cdot (TC)_n$ types.

What were the researchers to make of this? Could it be that a single-strand-specific endonuclease preferred to attack such sequences even if they were in the normal B form? Or was it that those sequences could accommodate, within the superhelical DNA, some new, until now undiscovered form of DNA?

This latter possibility was especially intriguing, since it would mean that we still do not know something very important about the DNA structure, and that this "something" could be of substantial importance to its functioning in eukaryotic cells. DNA structure specialists began to work on clarifying whether or not homopurine–homopyrimidine sequences form an alternative structure, and, if they do, to determine the kind of structure.

Some reasoned that these sequences may form a left-handed helix—not the Z form, but some other left-handed helix form, one that could be built with the help of molecular models. "No," others objected "the whole thing is that these are very homogeneous sequences, and so the two complementary strands may slip by in relation to one another, forming two single-stranded loops along the edges of the homopurine–homopyrimidine segment. The single-strand-specific endonuclease attacks these loops, which is precisely what makes such sequences hypersensitive." "It may be," still others insisted, "that such sequences form quadruple helices—yet another hypothetical structure proposed by theoreticians, based on playing with molecular models. The single-strand-specific endonuclease attacks the top of such a structure, where the single-strand loops must be."

The longer the question about the selective action of the single-strand-specific endonuclease enzyme itself remained open, the clearer it became that hypersensitivity to this enzyme does not lead to anywhere. Then, in 1985, Victor Lyamichev, Sergei Mirkin, and I (at that time, we were at the Institute of Molecular Genetics at Moscow) decided to apply the method of two-dimensional gel electrophoresis to the study of this involved problem.

The team constructed a plasmid carrying the $(GA)_{16} \cdot (TC)_{16}$ sequence. The plasmid's topoisomers were subjected to separation by the two-dimensional electrophoresis method. On the electrophoresis patterns one observed characteristic discontinuities like those seen in cruciforms and the Z form. In the control experiments (utilizing a plasmid into which no $(GA)_{16} \cdot (TC)_{16}$ section had been built), no discontinuities were present. The spots, corresponding to topoisomers that follow the discontinuity, disappeared from the two-dimensional patterns following the treatment of the insert-containing plasmid by a single-strand-specific endonuclease. These experiments demonstrated that under the impact of negative supercoiling, a certain alternative structure is actually formed in the homopurine–homopyrimidine sequence.

A question was then raised: "Could it be that what is formed is one of the already-known alternative structures—a cruciform or the Z form—and that the sequence requirements are just not as restrictive as is generally assumed?"

The answer to this was no, because the discovered transition turned out to be greatly facilitated by addition of acid to the environment. Moreover, given sufficient acidity, the transition is observed in DNA, which is relaxed and not at all supercoiled! It became clear that at issue was some kind of utterly new structure, since none of the previously known structures was so sensitive to an acid environment. Since this mysterious structure could be stabilized by an acid (i.e., by hydrogen ions), we called it the H form.

We obtained two very important quantitative characteristics of the H form. First, it is the value of the mobility drop in gel electrophoresis, which attested to the fact that in the H form, the strands are not twisted in relation to each other (as is the case with cruciforms). Second, supercoiling, (during which the transition from the B form to the H form occurs), is dependent on the acid environment. The theoretical analysis of this dependence demonstrated that, in the H form, for every four base pairs of the $(GA)_{16} \cdot (TC)_{16}$ insert, there is one proton-attachment site. Our attempts to invent a structure meeting this requirement, however, were an exercise in futility.

While we were racking our brains over this riddle, a paper published several years earlier by A. R. Morgan and Jeromy Lee (University of Alberta, Canada) caught our attention. They studied synthetic DNA, the poly(CT)·poly(AG) complex. They found that, in an acid environment, a pair of such double helices unite into a triple complex consisting of two CT and one AG chains. The second AG strand, turns out to be "superfluous." The triple complex consists of triads whose structure is given in Figure 35. Of great significance to us is the fact that the CGC^+ triad is formed by trapping one proton from the environment. This explains why such a structure is formed in the absence of superhelical stress only in an acid environment. Based on these data, we assumed that DNA's H-form structure is as shown in Figure 36. It is clear that such a structure must be very sensitive to the single-strand-specific endonuclease, since half the purine strand of the insert is in a single-strand state.

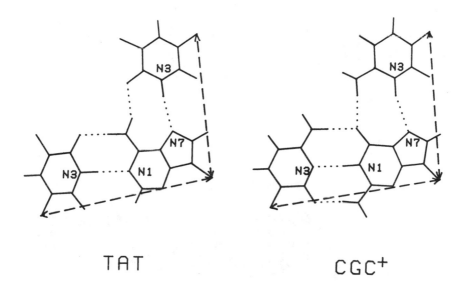

TAT   CGC⁺

Figure 35. The structure of TAT and CGC⁺ base triads that make up the DNA triple helix. Each triad is based on the ordinary Watson–Crick TA and CG pairs (see Figure 31), joined by a third base. This unorthodox way of joining bases was first observed by Hoogsteen in crystals. To form a GC Hoogsteen pair, the cytosine has to take over the extra hydrogen ion (i.e., proton) from the solution. The dotted lines indicate the so-called hydrogen bonds between bases. As in Watson–Crick pairs, both triads have the same distances between the nitrogen atoms bound with sugar.

We have also studied $(G)_n \cdot (C)_n$ inserts and other regular homopurine–homopyrimidine sequences. The H-form structure was observed in all cases. Can we then assume that the H form may arise in any homopurine–homopyrimidine sequence? According to our model (Figure 36), the homopurine–homopyrimidine character of the sequence is not enough. It is also necessary for the sequence to be a mirror-image palindrome (i.e., that it read the same from right to left and from left to right along the same strand—unlike the common palindromes forming cruciforms that read similarly along opposite strands). So it is clear why all regular homopurine–homopyrimidine sequences form the H form: They do so precisely because they belong to the mirror-image palindromes class. It is easy, however, to invent an irregular sequence—a mirror-image palindrome, say, like this one:

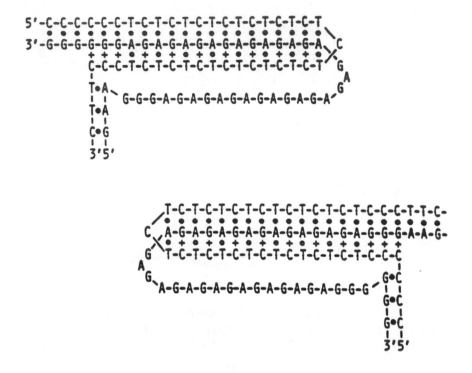

Figure 36. DNA's H-form structure for the (GA)$_{16}$ (TC)$_{16}$ tract, built into plasmid DNA. The structure's principal element is a triple helix consisting of triads shown in Figure 35. Two possible isomeric variants of the structure are shown. The Watson-Crick pairing is designated by filled circles, and the GC Hoogsteen pairing, involving the presence of an extra proton, is designated by plus symbols.

AAGGGAGAAGGGGGTATAGGGGGAAGAGGGAA

TTCCCTCTTCCCCCATATCCCCCTTCTCCCTT

The fact that the central TATA sequence is not homopurine–homopyrimidine is immaterial, because the central part of the sequence forms a loop in the H form (Figure 36).

Experiments showed that in mirror-image palindromes, the H form is formed much more easily than in sequences with even a slight deviation from the mirror symmetry. These and numerous other studies demonstrated, beyond any doubt, that the structure of the H form we proposed (Figure 36) is actually correct.

Thus, we now know that negative supercoiling can cause the formation in DNA of alternative structures of three types: cruciforms, the Z form, and the H form. It should be admitted, however, that in spite of numerous studies, the question regarding the possible biological significance of these unusual DNA structures remains obscure.

CHAPTER 12

Frontier Prospecting

ooooo

> . . . we have learned much about the cell's life and evolution, albeit
> not enough about how to prevent cancer. Rather on the contrary: we
> have seen the diversity of factors and mechanisms inducing cancer,
> which undermines the hope for universal therapy. And so these
> words by Ecclesiastes come to mind: " . . . in much wisdom is
> much grief: and he that increaseth knowledge increaseth sorrow."
> But scientists do not give up.
>
> —ROMAN B. KHESIN,
> GENOME'S INSTABILITY, 1984

Protein Engineering

For forty plus years now, the DNA molecule has served as a source of inspira-
tion for biologists and physicists, chemists and mathematicians, pharmacists
and physicians. It is possible that no other creation of nature has generated
such enormous interest. Among other things, this is due to the fact that the
DNA molecule has been and remains a magnificent example of how major
biological functions can stem directly from the structure of the molecule.

Strictly speaking, the question of the nexus between the structure of a
substance, on the one hand, and its functions and properties, on the other, is
a vital problem for many fields of science and technology. For example, in
some cases, one has to know what to use and how to use it in order to create
a material that will be superconductive and preserve its superconductivity in

possibly higher temperatures. In others, one must be able to create a poly-mer material that will be both strong and elastic. In still others, one needs a molecule that will kill cancer cells while sparing healthy ones.

Meeting all these conflicting demands is very difficult indeed. Like digging for diamonds when one is rewarded only occasionally, researchers are painstakingly amassing, bit by bit, information on the correlation be-tween structure and function. Against this backdrop, the DNA model of Watson and Crick looks like a diamond as big as the Ritz.

Regrettably, nature does not often treat us to such precious presents. Of all the variety of biological structures, it chose to provide only DNA with such a clear-cut structure. With protein, things are much more complex. Its structure has none of DNA's uniformity, and the link between an amino acid sequence and the protein spatial structure is hard to determine. Although many efforts have been made to solve this problem, protein has yet to reveal its secrets. (Incidentally, it is even more difficult to establish a link between the structure of a protein and its biological functions.)

Chemically, all proteins are single-type molecules. A protein molecule is a polyamino acid chain (i.e., a polymer—or rather, a co-polymer—composed of twenty kinds of amino acid residues). However, the properties and functions of these chains are very diverse! Chemists burst with envy when they see how quickly, effectively, and accurately (and under the most ordinary conditions) enzymes induce reactions that cost them much time and effort. A chemist must first treat the preparation with sulfuric acid, then with alkali, then expose it to oppressive heat or freezing cold. In living nature, however, everything is done with a truly divine simplicity. Naturally, practical people have long realized that enzymes are good for making wine, tanning leather, and doing many other useful things. But people do this purely empirically. They simply take the ready-made enzyme systems and "exploit" some of their particular possibilities.

We must not forget that in a living cell an enzyme is but a link in a very long chain of chemical events and that it is not necessary for it to display fervent zeal in accomplishing its limited objective. If it "overfulfills its quota," this will deplete the substrate and create a surfeit of final product. In general, in a good "cell economy," everything is harmoniously adjusted and there is no drive for "gross targets." However, when we get down to using enzymes for practical purposes, we would of course require them to not be overly selective vis-à-vis the substrate, or we would strongly require them to display other abilities. But the enzymes cannot meet out needs! At this point, a daring thought may come to mind to try to "improve" nature or, to be more precise, to modify natural enzymes or (more daring still) design new amino acid chains, so as to enable them to perform the functions that we desire better than regular proteins! Thus arises an idea that is best described as protein engineering.

The idea is not altogether novel. During the 1950s, soon after the first amino acid sequences were decoded, scientists began to think about such engineering. In their laboratories, the first attempts were made to duplicate nature and chemically synthesize some desired polyamino acid chains. Robert Merrifield at Rockefeller University scored the most success. He devised very successful methods of synthesizing polyamino acid chains (for which he was awarded a Nobel Prize in chemistry in 1984). Merrifield learned to synthesize short peptides, including many hormones. One merit of his technique was that all synthesis operations could be automated. Merrifield built an automatic machine, a "chemical robot," that, according to his intention, was to produce artificial proteins. In that pre-engineering era in genetics, Merrifield's robot caused quite a sensation by opening the way toward protein engineering. Soon, however, it became clear that the robot's products could not compete with those of nature.

First of all, the robot could not even remotely match the accuracy of production of amino acid sequences displayed by the humblest of cells. To put it bluntly, the robot erred in the most shameful fashion. One chain synthesized by it would have one sequence and another would have a somewhat different one. In a cell, however, all molecules of a particular protein are identical, all with the same sequences.

To make matters worse, molecules that were correctly synthesized by the robot refused to assume the spatial structure that is indispensable for the functioning of the enzyme. This defect was not confined solely to sequences invented for the sake of the experiment. The same fate also befell protein chains synthesized in strict accordance with the "prescription" of natural proteins that form a perfect spatial structure within a cell.

Then a sensational hypothesis, later confirmed through experimentation, was put forth: What mattered to the protein was not only its amino acid text but also the entire history of its emergence. It transpired, for instance, that at times, a longer predecessor protein is first synthesized in the cell and that it then doffs its redundant "tail."

It also became apparant that the cell had good reason to provide itself with ribosomes, tRNA, and the rest of the cumbersome protein synthesis apparatus. Any attempts to synthesize protein by purely chemical means have met with limited success.

More recently, researchers have discovered what is most likely the principal cause of this failure. It was found that the cell contains some specialized large proteins, dubbed *chaperons*, which help the polyamino acid chains, forming on ribosomes, to fold into a correct spatial structure. The word *chaperon* is French, and means a matron who accompanies one or more virgins for propriety and protection (in Spanish they are known as *dueñas*). The chaperon proteins perform a somewhat similar function, ensuring the polyamino acid chain's "growth" into a "mature" protein. The underlying mechanism of this fascinating process is now being subjected to a detailed study in many of the world's laboratories.

Be that as it may, nothing came out of protein engineering based on direct chemical synthesis. The only thing left to do was to learn diligently from nature (i.e., to search for required protein modifications that occur in nature), for in nature constant mutations are at work, leading to changes in amino acid sequences of proteins.

If we select mutants with desired properties—say, those capable of processing particular substrates more effectively—we can try to isolate from a mutant a modified enzyme to which the cell owes a new property. This has been the traditional approach of breeders from time immemorial. Geneticists have learned to map mutations and speed up their appearance by various techniques (e.g., radiation or chemical agents), but the principle has remained unchanged. This is the same "blind play" principle that, if one is to believe Darwin and his present-day followers, was used by nature to create the living world around us. But then nature had all the time in the world. With bacteria and viruses, which multiply very rapidly, the technique does yield some results, but it is extremely ineffective with enzymes of higher organisms. Still, until recently it was the only way of obtaining enzymes with new desired qualities.

With the advent of genetic engineering everything changed. Its methods have made it possible to create artificial genes with any desired nucleotide sequence. The genes are built into special DNA molecules—vectors—and the DNA is introduced into bacteria or yeast where an RNA copy is made of the artificial gene and a protein is eventually produced in strict accordance with the instruction contained in the artificial gene. Any errors in the protein synthesis are ruled out. Cell chaperons would take care of these alien proteins as if they were native. The principal objective is to select the desired DNA sequence; then the cell's enzyme system does the job impeccably by itself.

In this way, genetic engineering ushers in protein engineering in its most drastic form. Enzymologists (specialists who study the behavior of enzymes) have, in fact, found themselves completely unprepared for the unprecedented opportunities that suddenly opened before them. What would be rich is if it were possible for them to say, "Build this amino acid sequence and it will give you a hitherto unseen enzyme with this set of properties." Gene engineers are already fed up with the numbing chore of obtaining particular natural genes. Regrettably, however, we still have too poor an idea of how enzymes work to be able to design them. We have plenty of building materials and labor eager to get down to work, but do not have the drawings.

It is really a pity, but what is to be done? We must moderate our appetite. For besides the radical, there is also a moderate idea of protein engineering that advocates making small adjustments in natural proteins in order to understand the principal elements of their structure and why they work the way they do. In general, this is like working with mutant proteins, the resemblance similar to being that between a computer and an abacus. Indeed, genetic engineering techniques permit replacing at will any amino acid residue in the protein and changing the entire protein sequence as desired.

The smart genetic engineers have devised very elegant ways of obtaining desired mutations. For instance, say we want to replace a particular amino acid residue in a protein of our choosing. Let us assume that there are very good reasons to do this (or not so good, for now it has become immaterial). Before the actual replacement, we have to prepare a DNA vector (plasmid or virus) with the gene of the protein we want built into it. Of course we also must know the gene's nucleotide sequence and the amino acid sequence of the protein encoded by it. The latter sequence is easily deduced from the former, based on the genetic code table.

With the help of the genetic code table, it is also easy to determine what minimal changes have to be made in the gene to enable it to encode not the original protein, but one changed in accordance to our wishes. Let us assume that it is enough to make one change. For example, instead of the original sequence

... ATGTATTCGGATTAGCCGGT ...

we want the sequence

$$\ldots \text{ATGTATTCGTATTAGCCGGT} \ldots,$$

That is, in the middle of the gene section we need to have G replaced with T. It may be necessary to change two or even three nucleotides, but this does not matter. In any case, we must make a local replacement in the gene, which consists of many hundreds of nucleotides.

Must we resynthesize the whole gene because of this trifle? Of course not. We must synthesize only its small fragment of thirteen nucleotides, a complementary section in the middle of which is nucleotide G slated for replacement. Central in our synthetic fragment will be A, which will be complementary to the new, changed nucleotide T, rather than the original G. Thus, we synthesize:

$$\text{ATAAGCATAATCG}$$

There you are. We can then mix the resulting fragment under a set of conditions in a solution with the single strand (circular) of the vector DNA that carries our gene. The DNA ring and our short fragment form a section of the Watson–Crick double helix consisting of twelve nucleotide pairs. The central nucleotide pair is "evicted" from the double helix, as the pair is formed by mutually noncomplementary nucleotides G and A:

$$
\begin{array}{c}
\text{G} \\
\ldots \text{ATGTATTCG ATTAGCCGGT} \ldots \\
\text{ATAAGC TAATCG} \\
\text{A}
\end{array}
$$

Added to the solution at this point is the enzyme, the DNA polymerase that, using the fragment stuck to the single-stranded ring as a primer, completes building the DNA into a full ring in strict accordance with the principle of complementarity.

The result is an almost normal circular vector DNA that may be introduced into a bacterial (or yeast) cell, as is ordinarily done in genetic engineering. The only difference between the DNA and the original vector lies in the noncomplementary pair of nucleotides G·A (i.e., the duplex of the vector DNA is somewhat less than perfect). At the first doubling of the resultant vector, together with the bacteria carrying it, each of the daughter DNA molecules will become a double helix in its entire length. But one of the daughter molecules carries the G·C pair in the site of interest to us (and is, thus, identical to the original, nonmutant vector); the other has a T·A pair in that place. This latter molecule will be the desired mutant vector, based on which we shall be able to obtain the mutant protein of interest to us.

In this way the multiplying microbes will form a mixture of cells carrying the original vector with a nonmutant gene, and the desired vector containing a mutant gene. Now we only have to select from the mixture those cells that contain the mutant gene.

Selection According to Genotype

Generally speaking, selection of mutants is a routine thing for geneticists. They have always selected mutants that are in some way different from the original beasts—the wild type, as they are called. The difference may be in the shape of colonies when bacteria are sowed onto a nutritive medium, or in the shape of wings or color of the eyes (as in the case of the *Drosophila* fruit fly which geneticists are so fond of). But there must, of course, be some obvious—or, as geneticists say, phenotypic—difference between the mutant and the wild type. Otherwise, how can you tell one from the other? In other words, mutants have always been selected by the phenotype. Although geneticists have long been certain that phenotypic differences are always conditioned by genotypic differences, they were strengthened in their conviction only recently, with the advent of genetic engineering, when they learned to read DNA texts (i.e., the nucleotide sequences of genes).

So the time came when the old way of selection of mutants by phenotype no longer satisfied geneticists. Frequently, mutation manifests itself only in adult individuals, but exists from the beginning, in the zygote. This is true in regard to many human genetic diseases. Such diseases often do not manifest themselves at the embryonic stage. Some of them, it is true, can be identified by studying the chromosome set under the microscope. Many diseases, however, are due to point mutations (i.e., single nucleotide replacements), and these cannot be discovered with the help of the microscope. Frequently, a child assumed to be perfectly healthy at birth turns out many years later to suffer from an incurable genetic defect.

Some diseases are, pure and simple, genetic time bombs. For instance, Huntington disease is the result of a dominant mutation that manifests itself phenotypically only in mature age (thirty to forty years), when the unsuspecting victim already has a family and children (and has thus transmitted the malignant gene to the next generation). Then, in a matter of several months, a normal and not-yet-old individual falls into a state of debility. Since the mutation is a dominant one, half of the children are also doomed.

There are also situations in which mutation does not manifest itself phenotypically at all, and in which it is still necessary to be able to identify the mutant. This happened in protein engineering where mutant and non-mutant bacteria differ only in the structure of the enzyme totally unnecessary for the life of the bacteria. Cells are not very particular about the type of proteins they have to produce. However, protein engineers may have strong feelings on the matter. In that seemingly homogeneous mixture of bacteria, they have to identify those producing the desired mutant protein.

As can be seen, both protein engineers and specialists in medical genetics are badly in need of a methodology permitting selection of mutants, based directly on genotype (i.e., on nucleotide sequence of the DNA). The method of selection by genotype is based on a technique known as Southern blotting, which was proposed by E.M. Southern of the University of Edinburgh, Scotland, in 1975. Cells are sowed in a petri dish to permit every individual cell to give rise to an isolated colony. Like the cells begetting them, the colonies will be of two types—mutant and nonmutant. The dish is then covered with a nitrocellulose filter that is very much like blotting paper. Some of the cells from each colony get stuck in the filter. The filter is then treated with different chemicals, which destroy the cells, peel proteins off the DNA, and separate the complementary DNA strands. This rather merciless filter treatment leaves behind, in lieu of living cells, some formless little heaps of molecules of which the cells consisted, with DNA molecules present in the form of single strands. Several such filters are prepared for each colony.

The synthetic fragment, the one that served as a primer for the DNA polymerase, is then added. In our case (see p. 163), it is the ATAAGCATAATCG fragment. However, now this fragment carries a radioactive label on its end.

At this point, it is important to find out whether the DNA perched on the filter will seize the labeled fragment at different temperatures. As mentioned in Chapter 3, the DNA double helix is destroyed ("melts") when heated. Thus, as temperature increases, the complementary strands inevitably diverge. The precise temperature at which this divergence takes place depends on how perfect the double helix is. So the set with the perfect double helix, in which all thirteen nucleotides form complementary pairs, will remain stable at higher temperatures; unlike the set with the noncomplementary pair in the middle, for which the strands will separate during such heating. That is why, embarking on a quest for mutant colonies, identical filters are washed with the labeled fragment at different temperatures.

The quest technique is quite simple. At a low temperature, all remnants of cells (both mutant and nonmutant) will seize the labeled fragment and thus blacken the film put on the filter after incubation with a labeled fragment. The growing temperature will reach a point where the nonmutant DNA will no longer be able to hold onto the labeled fragment, whereas with the mutant one, given the perfection of the duplex it forms, this is not the case (i.e.,it continues to hold onto the labeled fragment). Correspondingly, the film will not get black in the former instance, but will in the latter. These are precisely the conditions under which one can tell the mutant colonies from nonmutant ones.

Now we only have to put the film on the original Petri dish with living cells, of which a "mask" was made using the nitrocellulose filter. The colonies that are found under the exposed spots of the film contain the mutant genes. The next stage, if proteins are at issue, will be the usual genetic engineering routine.

How will all this wisdom help prevent genetic diseases? Let us consider sickle-cell anemia, one of the most widespread diseases of this kind. As stated in Chapter 2, this grave genetic defect is due to a point mutation in a gene encoding the β-chain of hemoglobin, and A is changed to T in the mutant gene. As a result of this substitution, instead of glutamic acid, valine turns out to be the sixth amino acid residue of the protein chain. The result is catastrophic, for the protein's structure and properties change, and it largely loses its ability to carry oxygen. Even the form of red blood cells (erythrocytes) changes—from round and pucklike to sickle shaped (hence the name of the disease). Mutation leading to sickle-cell anemia is recessive; it manifests itself phenotypically, if the child gets mutant genes from both parents. As with many other genetic diseases, sickle-cell anemia does not in any way manifest itself at the embryo stage.

If I had been told fifteen years ago that soon it would be possible to identify sickle-cell anemia at the DNA level, I would simply have laughed. This is now a routine procedure. Sickle-cell anemia is quite widespread among the African-American population. Let us assume cases of the disease in the families of both spouses, thus giving grounds for fear that both husband and wife carry mutant genes in the recessive state. Prior to genetic engineering, this couple would just have been told in the genetic counseling center that there was a 25 percent probability of their giving birth to a child inheriting mutant genes from both parents, and thus carrying the disease. These days, however, the would-be mother is placed under painstaking medical control. During the twelfth week of pregnancy (regrettably they cannot, as yet, do this earlier), the physicians take embryo cells from the liquid surrounding the embryo and send them to the research center for analysis.

There the multiplying embryo cells are sowed, like bacteria or yeast cells, and the test for the desired gene is conducted, as described earlier. However, in the process the labeled synthetic fragment of DNA complementary to the *normal* gene section of the β-chain of hemoglobin is used, since it is the normal gene that they are after (the mutation in question is recessive and it is, therefore, enough for the embryo to have one normal gene). If the test demonstrates that the embryo is endowed with a normal gene from at least one of the parents, then everything is in order and the child will be born healthy. If the embryo turns out to have no normal gene, the newborn will have sickle-cell anemia.

Thus, the question of whether or not the child is afflicted with sickle-cell anemia will be accurately answered in two weeks. In the event of an unfavorable diagnosis it will still be possible to interrupt pregnancy. There are already children who have passed such genetic diagnostics at the embryo stage. The molecular-genetic nature of many other known hereditary diseases is now being investigated at an accelerated pace. Who would have thought that pure basic research into DNA melting would yield such surprising practical benefits?

DNA and Cancer

We live in a fascinating time. Only yesterday we spoke of standing on the threshold of a new scientific and technological revolution, the threshold of a century of biotechnology. Before we could say "knife," however, we had stepped into that biotechnology era! Never before had scientific discoveries found practical applications as swiftly as they do now.

Before our eyes, the methods of molecular biology have penetrated such unlikely fields as archeology and criminology. Scientific journals and mass media are full of breathtaking reports about extracting and successfully cloning a DNA fragment from the mummified remains of an Egyptian boy who lived some 2,400 years ago or about the use of DNA, taken from four-year-old blood stains, in the positive identification of a criminal who left the stains.

There was once considerable talk of humankind facing the prospect of genetic degeneration. Now, however, scientists have developed, and in many cases applied, methods that permit them to determine at the embryonic stage whether or not the human embryo carries a particular malignant mutation. For many decades, enzymologists dreamed of consciously changing the chemical structure of enzymes in order to learn how to give them desired properties. This has become a reality, providing enzymologists with unprecedented opportunities to conduct basic research into the mechanism of enzyme reactions and to create artificial enzymes with desired properties.

Before the advent of genetic engineering, it was not possible, in practical terms, to even approach a study of the structure of the genes of higher organisms and a detailed comparison of genes in different cells of a multicellular organism. This possibility, now a reality, has already produced crucial discoveries that have shaken classical and molecular genetics to their very foundations. One example that comes to mind (as discussed in Chapter 7) is that of the dismemberment of genes and their notable restructurings and changes during the formation of specialized cells (i.e., during the process of differentiation). However, the chief target for researchers at present is cancer and the mechanism that triggers its development.

Cancer stands apart from other diseases because a cancer cell is an organism's own cell, but one that behaves like an alien intruder, a "fifth column," if you will. Until a certain time, the behavior of such a cell does not differ in any way from that of the rest, as long as it abides by the "rules of the game" accepted in a multicellular organism. In an adult organism, these "rules" state that the division of cells be strictly controlled in different ways for different tissues, and that for some (such as nerve cells), division be strictly forbidden. This cannot be otherwise, for if each cell were left free to divide as it pleased, the organism would turn rapidly into a formless mass of runaway cells.

At a certain time, however, a "law-abiding" member of the community of differentiated cells suddenly stops abiding by the rules of the game and goes berserk. It embarks on unrestrained growth—that is, it becomes a cancerous cell. In the process, the cell transmits this property to all its offspring. Eventually comes metastasis (literally, "beyond control")—the spreading, via the blood, of the dividing cancerous cells from the disease's original location to newer sites around the body. This all results from degeneration of a single cell.

The perfidy of the traitor cell lies in that the organism's immune system, acting as a "security force," still continues to regard the cell as one of its own: a friend, not a foe. Consequently, the organism, whose immune system is quite capable of turning the tables on almost all alien invaders—bacteria or viruses—very frequently feels helpless when faced with an "inside enemy."

Doesn't the organism have a "secret police" of its own, capable of uncovering and neutralizing the traitor cell? As a matter of fact, it has! The policing function is discharged by a certain class of T lymphocytes, T killers, to be exact. These can detect and neutralize cancer cells, or, simply put, kill them. Everybody knows that the secret police are not to be trifled with. Allowed sufficient freedom, they may engage in indiscriminate killing and exterminate the normal, "law-abiding" cells as well. Nor can one dispense with the police, for this will sharply increase the chances of cancer.

Steven Rosenberg, from the National Cancer Research Institute, devised a new technique for treating cancer, based on stimulating the organism's "secret police." For this purpose, T killers taken from a patient's blood were treated with a specialized protein, the growth factor of T lymphocytes. Called interleukin-2, the protein is produced in large quantities by standard genetic engineering techniques. The multiplied T killers were then introduced into the patient's blood. In this way, Rosenberg was able to cure completely a case of melanoma that used to be regarded as an incurable form of skin cancer. A drastic decrease in the size of tumors was observed in other patients, as well.

Rosenberg's studies aroused immense interest on the part of specialists and the general public. However, it will take years before it becomes clear how effective and universal this therapeutic method is. Let us not forget the words of the late Russian geneticist Roman Khesin, given in the epigraph to this chapter.

From where do the traitor cells come? Considering that their bad behavior is inheritable, the first thing that comes to mind is to assume that some change in the DNA of the cell in question is responsible, forcing a normal cell to go berserk. (Incidentally, this supposition, which would cause no objections in the case of bacteria—the reader will recall Avery's experiments, described in Chapter 1—would not be nearly as obvious with respect to cells of higher animals.) We know that the cells of a multicellular organism can often modify their behavior drastically even without changes in DNA. Thus, one single fertilized egg cell can give rise to a whole organism consisting of cells differing in their properties and functions (e.g., the cells of the liver and bones). Almost all of these cells, however, contain all the original genetic information.

In most cases, the cells' differentiation has to do with a change in gene activity: The genes themselves, and DNA sequences in general, remain unchanged; but in some cells of a multicellular organism, one group of genes operates whereas in other cells a different group does. Hence a coherent theory that cancer is simply cell dedifferentiation triggered by some internal causes. However, the theory has its own difficulties. The principal one became apparent at the very beginning of this century, during experiments on animals that showed that cancer could be induced from outside (e.g., by infecting animals with a virus). Viruses capable of inducing cancer in animals (of which there are many) are called oncogenic.

In the 1940s, the remarkable Russian scientist Lev Zilber put forward his virus–genetic theory—one of the most fertile ideas proposed over the lengthy history of cancer research. The theory can be summarized as follows. Penetrating a healthy cell, DNA of the oncogenic virus incorporates itself into cellular DNA, modifying the latter's genetic property and causing its uncontrolled division. The intruder virus replicates along with the cell's own DNA and is transmitted to future generations.

The virus theory gained proponents very slowly. Nobody, of course, took issue with the fact that some of the tumors observed in animals are virus induced. However, serious doubts lingered regarding the potential universality of the concept, including its applicability to human tumors, for it is well known that cancer can be induced by the most diverse agents—both physical and chemical. Various chemical compounds, called *carcinogens*, are known to sharply increase the likelihood of the development of a tumor. So what do viruses have to do with all this?

A crucial blow to the virus theory was dealt by the discovery that RNA, not DNA, serves as genetic material for many oncogenic viruses, and RNA cannot incorporate itself into DNA. What, then, allows the virus to incorporate genetic information into DNA? The fact that oncogenenic viruses often carry RNA rather than DNA was found long before RNA's ability to synthesize DNA needed for incorporation had been discovered. Thus, at that time, it looked as if the changes leading to the development of cancer could occur without DNA; consequently, the virus–genetic theory could not stand on its feet.

Some of the biologists, however, were quite reluctant to relinquish Zilber's idea, which still impressed them as simple and concrete; experiments continued to provide evidence that oncogenic viruses can induce cancer. While this was clearly at variance with the notions of molecular biology at that time, the search for the mechanism enabling RNA viruses to transmit their genetic information to the cell continued. Howard Temin displayed a special pertinacity that was eventually rewarded. In 1970, Temin and David Baltimore found an enzyme in RNA-containing oncogenic viruses, named reverse transcriptase, that synthesizes DNA on viral RNA once the virus is inside the cell. This "viral" DNA is incorporated into DNA of the cell, thus causing malignancy.

This discovery (as has already been mentioned in Chapter 4), was quite a milestone for molecular biology and proved to vindicate the virus–genetic theory of cancer beyond a doubt. The effects of carcinogens and many of the discrepancies in the virus theory had receded into the background. The next goal was to isolate viruses corresponding to the different types of cancer and to learn how to control them.

Time passed, however, without any tangible accomplishments. One stumbling block was the failure to discover cancer-inducing viruses in human beings. In general, it is not surprising that research on animal cancer has significantly outpaced research on human cancer. A virus could, of course, be obtained from the blood or from a surgically removed human tumor. But how could you be sure that it was really a cancer virus? Would you infect a healthy person to verify that it was a cancer virus? This difficulty, it is true, has been obviated, but only partly.

Quite a long time ago, biologists learned to cultivate cells of humans and animals in vitro (i.e., outside the living organism). Compared to bacterial or yeast cells, these are incomparably more difficult to cultivate, but they permit experiments that would otherwise be impossible. Moreover, differentiated cells in vitro ordinarily display a "civilized behavior" by obeying the rules they were schooled to abide by in a multicellular organism. At the flat bottom of a glass vessel, for instance, they form only one layer and then their growth stops. This, however, is not the case with cancerous cells! As they divide, they swell beyond the confines of the monolayer, forming a compact outgrowth that is quite visible through a microscope. This means that a cancerous degeneration of cells can be successfully studied in vitro.

Eventually, Robert Gallo from the National Cancer Institute was successful. He obtained an oncogenic virus responsible for a form of leukemia in humans. (Incidentally, when the HIV was discovered later, the two viruses turned out to be close relatives.) However, only rare forms of human cancer have a viral origin. The absolute majority of cancer cases are not in any way related to viruses.

Animal cancer research fared no better, despite the embarrassing ubiquity of oncogenic viruses. It appeared that in most cases, viral DNA is built into the animal's DNA at birth. Why, then, do all animals not suffer from cancer right from birth? It was then assumed that, in addition to the presence of viral DNA, the cell needed something else, a sort of command to trigger cancer. Is a carcinogen the signal that spurs viral DNA into action?

This would be tantamount to stating that the carcinogen is precisely the chief agent that induces cancer. If viral DNA is always present in the cell from the outset, why all the talk about viruses? Could it be that DNA of a given animal is organized this way, whereas cancerous growth is triggered by a carcinogen? It may be that the carcinogen activates either the installed viral DNA or other DNA sections. Or, magnanimously leaving DNA alone, the carcinogen may initiate a so-far unknown signal for differentiation, following which the cell "forgets" the "rules of the game."

Thus, in less than a decade after the triumph of the virus theory of cancer, researchers were back to square one. The wheel had come full circle, and there was no getting away from the accursed differentiation problem. As always, however, a glimmer of hope still lingered. What if carcinogens do affect DNA (whether the virus or one of the other sections is immaterial), thereby changing DNA's text? To put it differently, what if carcinogens are actually mutagens?

The carcinogen problem has long attracted the attention of researchers, and not only because of their theoretical investigation into the origins of cancer. The fact is that each new chemical compound human beings are exposed to must be tested for carcinogenicity. There are many examples of how thoughtlessness or negligence in such a serious matter proved akin to planting a time bomb that eventually proved fatal.

How does one determine whether a particular material is or is not carcinogenic? For many years, attempts have been made to devise fast and reasonably cheap techniques for testing chemical substances for carcinogenicity. (Incidentally, this testing still remains the most costly and time-consuming procedure in putting any new drug on the market.) It is still considered necessary to subject test animals to the effects of the new preparation and to compare their behavior with that of the control animals up to the time they die a natural (or unnatural, if the drug happens to be carcinogenic) death. Couldn't this procedure be simplified?

A vast body of findings, amassed by Bruce Ames (University of California at Berkeley) through a labor-intensive testing of drugs for carcinogenicity, enabled him to devise a highly effective testing technique. In 1975, Ames proposed that substances be tested for mutagenicity rather than carcinogenicity. In mutagenic testing, you can dispense with animals or even with their cell cultures. You can simply use bacteria, for which the techniques have long existed, for rapidly calculating the rate of mutation (i.e., change in the DNA text). Ames proceeded to refine these techniques, checking in the process the hypothesis that mutagenicity and carcinogenicity actually boil down to the same thing.

It would seem that for the purpose of testing one has to take chemical compounds known to be carcinogens and test them for mutagenic effects. As they say, however, there is many a slip between the cup and the lip. The fact is that circulating in the blood of the organism, a carcinogenic compound undergoes chemical changes. The liver, for example, is literally crammed with enzymes capable of inducing all sorts of modifications. So it may well be (as has been proven in some cases beyond a doubt) that cancer is induced not by the original substances, but by the products of their metabolism once inside the organism. For this reason, before he tested substances for their mutagenic effects, Ames processed them with an extract from animal liver.

Ames tested 300 substances for mutagenicity, including "inveterate" carcinogens as well as quite innocuous substances. His tests showed a clear and unmistakable correlation between carcinogenicity and mutagenicity. In 90 cases out of 100, carcinogens also proved to be potent mutagens; at the same time, only 13 percent of noncarcinogens exerted a mutagenic effect.

These results are highly convincing. They show that Ames's test is effective, at least for mass-scale testing of chemical compounds. Judge for yourselves: Ames, aided by only one assistant, was able to test 300 substances in a short time! This would have taken many people decades of painstaking work, if conventional methods were used to accumulate findings on the carcinogenicity of all these substances.

Ames had set himself a purely practical goal—to develop an effective and cheap carcinogenicity test. His findings, however, also proved of great importance for understanding the nature of cancer. Actually, they left no doubt that carcinogens cause cancer precisely by changing DNA of the cell (i.e., by causing mutations).

Ames's clear message was that the original events that eventually lead to cancer evolve in DNA (i.e., in the genetic material). This being the case, reinspired molecular biologists rushed forward again to attack the problem of cancer. This time, however, thanks to the decade that had elapsed since the appearance of the works by Temin and Baltimore, they were already well equipped with genetic engineering's powerful DNA-manipulation techniques.

In 1979, experiments were done that proved, once and for all, the genetic, DNA nature of cancer. Conducted on mice, the experiments, however, did not differ in principle from those on transformations in pneumococci done by Avery forty years before. Robert Weinberg (Massachusetts Institute of Technology), the author of the experiment, reasoned in this manner: from Ames's experiments it follows that carcinogens must affect some change in DNA, after which it acquires the ability to turn a normal cell into a cancerous one. If that is really so, then by isolating DNA from cancerous cells and transferring it to healthy ones, we should (allowing, naturally, for a certain rate of probability, as in any transformation) observe the transformation of normal cells into cancerous ones.

Weinberg isolated DNA from mice tumor cells whose degeneration had been induced by a powerful carcinogen. He then conducted transformation experiments. Cancerous DNA was added to a culture of healthy mouse cells, known under the code name of NIH3T3. The results of the experiment were as follows. In five cases out of fifteen, the NIH3T3 turned cancerous. On the other hand, no malignant degeneration occurred in any of the ten control experiments in which normal DNA had been added to the NIH3T3 culture.

The properties the of cells that had been degenerated by the transformation technique were then tested on mice: Cancerous NIH3T3 cells were implanted in healthy mice, which developed malignant tumors.This is not the whole story, however: A degeneration of NIH3T3 cells into cancerous ones was also achieved with the implantation of DNA isolated from *human* cancerous cells! DNA taken from healthy human tissues, however, did not lead to a malignant degeneration of NIH3T3 cells.

At this stage, it was time for genetic engineers to join in the effort. Since human DNA causes transformations, this means it carries an oncogene (i.e., a section responsible for the deadly changes). The hunt for oncogenes began. Oncogenic viruses proved to be very useful in the search for oncogenes. They turned out to possess ready-made oncogenes. In a brief time, about thirty oncogenes were cloned and described in detail, including their complete nucleotide sequence. Specialists believe that this relatively small set of genes is actually responsible for the entire variety of cancerous diseases in animals and human beings.

Each oncogene turned out to have a cellular "brother," a normal gene called a *protooncogene*. From the molecular–genetics viewpoint, oncogenes, like protooncogenes, are ordinary structural genes (i.e., each carrying information about the structure of a particular protein). A protooncogene per se is not dangerous. Moreover, protein products of protooncogenes play a crucial role in inter- and intracellular communication.

Indeed, to demonstrate exemplary behavior in a harmonious family of multicellular organism cells, each cell has to obey the incoming signals. One crucial signal, for example, tells the cells to multiply (divide). If, for instance, you cut yourself while shaving, then the skin cells around the cut will begin to divide intensely to cure the wound. The messengers bringing the cell the order to divide are specialized protein molecules—the growth factors. They deliver the "messages" to other protein molecules, receptors built into the cell's external wall.

Thus, the cell has received the message: The growth factor got into contact with its receptor. But everything in the cell obeys DNA hidden inside the nucleus. To be heard, the signal must also penetrate the cell's external wall, cytoplasm, and nuclear wall. The signal gets transformed several times on this complex itinerary, transferred by specialized intracellular couriers, with a number of proteins taking part in the process.

A crucial fact is that proteins—products of protooncogenes—are growth factors, receptors, and other proteins of inter- and intracellular communication. In what way does a malignant oncogene differ from a benign and very indispensable protooncogene?

We now know of a number of mechanisms that turn a protooncogene into an oncogene. It may simply be a point mutation—replacement of one amino acid residue. It may also be a chromosome restructuring that results in the protooncogene's transfer to another chromosome. One attendant feature of the process is the upset regulation of synthesis of the normal protooncogene product, or a truncation of the gene itself in the restructuring process. Another mechanism is that the protooncogene itself will remain in its place while the regulatory region from another chromosome gets shifted to it.

From detective stories and movies we know that the deadliest of spies is the one who has secured for himself access into the enemy army's chain of command so that he can feed a false attack order down the line. This is precisely how oncogenes behave. By an accelerated outgrowth of the growth factor, producing a defective receptor or some protein of intracellular communication, the oncogene forces the cell's DNA to comply with the false signal to divide. Oncogene-carrying cells begin to divide in an uncontrolled fashion, with the daughter cells also carrying oncogenes (i.e., being given the division signal). This is how cancer starts.

Now a new stage in the struggle against the disease has begun. Indeed, after oncogene products have been thoroughly investigated, it will certainly be possible to invent a way to render the proteins harmless. Of course, the most radical solution would be to damage or "repair" the oncogene, and we shall discuss such a strategy later. Another possibility is damaging proteins with the help of antibodies.

Researchers working with oncogenes in dozens of laboratories around the world are quite optimistic. According to M. Wiegler, one of the researchers who isolated oncogenes, the end of the journey may be in sight. (Wiegler is working at Cold Spring Harbor Laboratory which, having been formed on the basis of the Phage Group, had for a long time been headed by Delbrück. Then, for many years, the laboratory was directed by Watson, who has made the study of the nature of cancer at the DNA level the chief thrust of his laboratory's research effort.) Ordinarily, such optimistic statements are received with skepticism. Too many times it seems that we are but one step from understanding the nature of cancer, only to be bitterly frustrated. At present, however, there are certainly more reasons for being optimistic than ever before. This is because the study of the problem of cancer is now being firmly pursued at the level of DNA.

Program of Death

One of the most severe psychiatric disorders, so-called suicidal syndrome, consists in the patient's persistent and virtually uninhibitable drive to self-kill. Unfortunately, in many cases, in spite of all efforts of relatives and friends to prevent the fatal ending, a patient succeeds in reaching this macabre goal. Something terribly wrong happens in the patient's brain that transforms the natural instinct of self-protection into a totally unnatural desire to self-destruct. A patient behaves as if he has been programmed to kill himself.

Beyond a doubt, from the viewpoint of an individual, the suicidal syndrome is most unnatural behavior. But what about the population, the society in general? Would it not be nice if Hitler and Stalin had suffered from this syndrome and killed themselves instead of slashing the world population by millions and millions of decent human beings? It is not unreasonable to believe that our world would be a much safer place if some individuals were never born or, if born, killed themselves before inflicting misery upon other members of society.

Genetically programmed, complete or partial self-destruction is often met in living nature. To give life to the next generation, a salmon, obeying its genetic program, leaves the spaciousness of the ocean, enters small rivers and springs, and crawls up to their very sources while overcoming tremendous obstacles. All along this path of death, the salmon does not eat anything, using only resources that were collected in the ocean. After discharging the roe or sperm, the fish dies of hunger. Thus, organisms are genetically programmed to sacrifice their lives for the sake of safe birth of future generations.

We also witness the massive "temporary suicide" every fall when the leaves fall off of deciduous trees in a strictly genetic-programmed manner. Additionally, it was recently discovered that the DNA of each cell of eukaryotic, multicellular organisms carries a special program that, if turned on, initiates a series of events leading to cell death. This programmed cell self-destruction was dubbed apoptosis, which means "a falling off" in Greek.

Understanding of apoptosis and its role in nature is definitely the most important development in the field of molecular and cell biology in recent years. It is amazing that such basic, widespread, and important phenomenon had been essentially overlooked for so long. However, very recently it has become the focus of attention of a huge army of researchers and physicians. The "cell suicidal syndrome" more than deserves such interest. Indeed, if cells are equipped with necessary tools to kill themselves and only need a proper signal to do so, all we need to learn is how to send these signals to "nasty" cells that we want to rid the body of—that is of course, cancerous cells.

It appeared that one of these signal proteins, which sends the death command to tumor cells, has been known for a long time. The protein is known as tumor necrosis factor (TNF). Like other signal proteins, it binds to special receptors on the surface of the cells, and this binding initiates a long relay of events leading down to DNA. As a result, special apoptosis genes are switched on, and new enzymes are expressed, among them proteases and nucleases that digest cellular proteins and nucleic acids. The cell is degraded into pieces, and the pieces are swallowed by special cells, macrophages, our body's janitors. No traces remain of the cell that had received the death command.

"Fantastic," the reader thinks, "is this not the ultimate cure of cancer? TNF is specific to tumor cells so normal cells will not receive the death command." Beyond a doubt, TNF is a very promising tool in the fight against cancer. However, very soon after its discovery two decades ago, it became clear that TNF does not kill most cancer cells. Normal function of TNF probably consists of eliminating tumor embryos consisting of very few cancerous cells. At the same time, the organism needs to protect itself from cases, even very rare ones, when the TNF death command is mistakenly received by normal cells. Therefore, as was found recently, in conjunction with turning on the process of apoptosis, TNF induces the production of a protein which turns on the genes combating cell death. This protein is known as nuclear factor kappa B (NF–κB).

The NF–κB protein is viewed as one of several factors that prevent cells from suicidal behavior. Experiments performed by Baltimore's group at MIT, with so-called knockout mice, in which the gene coding NF–κB was specifically destroyed, demonstrated that the lack of NF–κB leads to such a massive suicide of mice liver cells, that the animals die before birth.

Thus, yet another possibility in the fight against cancer is being tried. One can find drugs that inhibit NF–κB, thus making cancer cells much more compliant to the death command from the TNF protein. But the inhibition of NF–κB may appear even more promising. Apoptosis is known to be triggered by other means than TNF. If cells (specifically, the DNA molecule) are severely damaged, they may choose to commit suicide (just like numerous incurably ill clients of the retired pathologist, Jack Kevorkian, do). However, NF–κB normally prevents them from doing so. Without NF–κB, tumor cells damaged by radio- or chemotherapy will finish the job themselves through the process of apoptosis.

The decision to commit suicide in the case of excessive DNA damage is made by another protein, named p53. One should not be misled by the modest code name of this protein: there is no other protein that currently attracts as much attention from researchers and medical doctors as p53. This protein controls DNA damage during numerous cell divisions in the course of the development of a multicellular organism. We already know that DNA damage, somatic mutations, and other DNA transformations are the major causes of cancer. An outstanding function of p53 is that if it finds that the DNA in a cell is significantly damaged, it has the power to sentence the cell to death—that is, it triggers the process of apoptosis. If p53 is inactivated because of a mutation or due to other mechanisms, the DNA damage cannot be properly controlled and numerous forms of cancer can emerge. According to the current view, inactivation of p53 is the primary cause of cancer in the absolute majority of cases. For instance, when researchers revisited a urine sample, stored since 1967, from the famous 1960s politician, Hubert Humphrey (who died of bladder cancer in 1976), they found that cells in the urine contained mutated p53. In about 60 percent of cancer cases studied, p53 proved to be inactivated. As a result, tremendous efforts are currently being made to find ways to keep p53 active.

Recently, Swiss researchers have found that in some of the most vicious forms of cancer, like melanoma, the cancer cells send signals to T killers to commit suicide. As previously discussed in Chapter 7 and an earlier section of this chapter, T killers are immune system cells that normally recognize and kill cancerous cells. It is as if criminals, using police radio frequencies, sent the police a command to kill themselves, and the police obeyed the command and commited suicide! Although this may happen only in a weird Hollywood thriller, it is exactly what happens within the human body in the case of melanoma. The body's major police force—T killers—commit suicide, obeying the vicious command from the most dangerous of murderers, melanoma cells. This stunning discovery of Swiss researchers explains why our immune system is so helpless in cases of the most terrible forms of cancer. At the same time, it adds to the feeling, which many researchers share, that understanding the process of apoptosis and its relation to cancer will eventually provide doctors with new powerful tools to fight the disease.

Of course, it remains to be seen whether these newly emerging and extremely promising strategies to fight cancer will really work. The study of apoptosis may well become just another fascinating piece of our knowledge "about the cell's life and evolution," rather than a powerful weapon in our war against cancer (see again Roman Khesin's prophetic words in the epigraph to this chapter). Even the accumulation of enormous volumes of information about oncogenes has yet to lead to a real breakthrough in cancer therapies. And still, while typing these lines, I feel more strongly than ever that this time everything will be different and that the unraveling of the genetically programmed ability of the cell to kill itself will eventually lead to a real breakthrough in our protracted battle with cancer.

DNA and the Heart

DNA, the "most important molecule," has many faces. It delights people by its "well-knit" structure and holds out the promise of revolutionizing medicine and farming. But, like a pagan deity, it can also give people an almost mystic sense of terror. Just think of it: a slight change in its chemical structure can turn it from boon to blight. A person may go about his or her business, unsuspecting, blithely giving in to earthly joys, but already carrying the malignant gene that will bring death in the prime of life. Unlike the mythical gods, DNA does not know compassion or justice. Under its inexorable laws, the ill fate that befell the father may also threaten the son.

Whereas in the past (and often now, for that matter) people were convinced that all their troubles were sent by God or caused by an evil eye or their sins, now they are having to live with the idea that their troubles are caused by DNA, by genes. This is true not only of cancer and sickle-cell anemia but of atherosclerosis, for a damaged gene plays a crucial role here, as well. This is especially true in the early form of the disease.

Early atherosclerosis afflicts people with a damaged gene responsible for the structure of the specialized receptor that sits on the surface of cells, mainly liver cells. These receptors identify the special fat particles circulating in blood, the so-called low-density lipoproteins (LDL), which are literally crammed with cholesterol molecules. Binding with their receptors, LDL corpuscles get inside cells, where cholesterol delivered by them is utilized for the production of hormones and for other purposes. A damaged gene produces a deficient receptor incapable of binding LDL. This leads to an LDL accumulation in blood and to a sharp increase in the level of cholesterol. Cholesterol begins to settle on blood vessel walls, which leads to atherosclerosis. A patient whose cells have no LDL receptors at all dies of complications caused by atherosclerosis (usually a heart attack) sometimes before the age of twenty.

Happily, this rarely happens (one person per million). Indeed, for this to occur, both LDL receptor genes have to be damaged (i.e., paternal and maternal). As geneticists say, the person must be homozygotic by the damaged gene.

Seen much more frequently (i.e., one in 500) are people who are heterozygotic by the LDL receptor gene, that is those who received a damaged gene from only one parent. In this case, the good gene ensures half the normal number of good LDL receptors, and, as a rule, atherosclerosis develops by the age of thirty-five.

Although heart attacks are not caused solely by genes, it was precisely the investigations into early, hereditary atherosclerosis that led to the discovery of LDL receptors and their key role in regulating the level of cholesterol in blood. This, in turn, finally proved that cholesterol level is the decisive factor in the development of the disease.

Thus, a new, scientific basis for the search into ways to prevent atherosclerosis had been provided. One way is to increase the number of LDL receptors. There are already very efficient drugs (they are known by different names: lovastatin, mevacor, zocor, atorvastatin, lipitor) that make this possible (only, of course, if the patient has at least one good gene for the LDL receptors). These drugs are decreasing the cholesterol level in millions of patients worldwide, sharply diminishing their chances of dying of a heart attack.

Another way to decrease the level of cholesterol, which is accessible to everyone, is to lessen the amount of cholesterol getting into the blood (i.e., to cut down on the amount of LDL itself). The most cholesterol-rich product of daily consumption is the yolk of an egg. Those whose families have a record of atherosclerosis (and there is hardly a family nowadays that has been spared by the disease) would do well to avoid egg yolks. This would mean bidding farewell to eggs for breakfast. Even a glass of milk is not altogether innocuous.

These discoveries, which have changed both the lifestyle and life expectancy of millions of people in America and elsewhere in the world, were made by Michael Brown, a biochemist, and Joseph Goldstein, a medical geneticist (Texas Southwestern Medical Center), for which they received a Nobel Prize in 1985.

Changing Heredity

Our knowledge of the genetic nature of different diseases grows at an accelerated pace. We have concentrated briefly on two of the most common diseases—cancer and cardiovascular disorders—whose genetic connection has been established beyond doubt. There is also growing evidence that the principal psychic disorders, such as schizophrenia, also have a genetic origin. As we discover the genetic nature of more and more diseases and proceed to decode at a faster pace the full nucleotide sequence of the human genome (thanks in part to the successful implementation of the Human Genome Project), there arises an acute need to develop methods to modify the genome purposefully in a way that would "silence" malignant genes, such as oncogenes, and at the same time give the good genes, such as the genes of LDL receptors, the possibility "to speak up." Increasing numbers of research laboratories in both universities and industry are being drawn into the search for such approaches.

What ideas and accomplishments are there in this field? The first strategy of purposeful modification of gene expression is to block a gene's products at the translation level (i.e., at the level of synthesis of protein from the RNA template). An extra DNA section is introduced into a cell (either in the form of a plasmid or built into the genome). The DNA section is structured as follows. It carries a strong promoter, followed by a sequence such that its RNA replica will be complementary to the part of mRNA of the gene that must be "silenced." This complementary RNA is called an antisense RNA. Thus, this particular strategy is known as an antisense-RNA strategy.

Introduction of an extra DNA sequence creates in the cell a constant high concentration of antisense RNA, so that the entire RNA, synthesized on the selected gene, turns out to be bound with the antisense RNA and cannot normally be translated on a ribosome. The upshot is that the protein either fails to be synthesized at all or the part that is synthesized is one that cannot function.

The antisense-RNA strategy is already widely employed to block particular genes. For example, the strategy helped produce a new variety of "never-get-old" tomatoes that cease developing at the green stage and never grow ripe. Built into the genome of this remarkable vegetable was a DNA section that synthesizes antisense RNA that blocks the synthesis of one of the enzymes producing the tomato's ripening hormone, which is the ordinary ethylene molecule. Thus, if we put such a "non-ripening" tomato in an atmosphere containing a small quantity of ethylene, it will rapidly ripen and turn red. It is clear that this new tomato variety is very convenient to transport and store. Such tomatoes are already being sold in some supermarkets.

In the preceding form, the antisense strategy obviously cannot be applied to humans in order to specifically silence wrong genes. To achieve this goal, a piece of single-stranded DNA (an oligonucleotide) is delivered from the outside, as a drug. The oligonuclotide sequence is complementary to a site on mRNA, corresponding to the gene to be silenced. Therefore, instead of antisense RNA, the antisense DNA is used. There are several reasons why DNA is a much more promising molecule as an antisense drug than RNA: DNA is chemically more stable than RNA so it is easy to store, and it lives longer inside the cell, although it is also degraded. To prevent quick degradation, the antisense oligonucleotide is disguised from cellular enzymes by special chemical modification. However, the major reason why DNA oligonucleotide is very efficient in the antisense strategy is as follows.

When the antisense DNA oligonucleotide finds its target and forms a short duplex with mRNA, a special enzyme, called RNase H, attacks the duplex and degrades its RNA part. This is a specialty of this enzyme: It recognizes DNA/RNA heteroduplexes and digests their RNA part. As a result, the synthesis of bad protein proves to be prevented even before mRNA reaches a ribosome. This is very fortunate because a long oligonucleotide cannot be used as drugs, whereas binding of a short oligonucleotide with mRNA cannot prevent a ribosome from doing its job. In the case of tomatoes and similar situations, antisense RNA is very long, so it covers a significant portion of mRNA thus preventing the ribosome from reading it. Antisense drugs would never work without RNase H.

Several antisense drugs designed to shut down various oncogenes have shown remarkable ability to slow down the growth of tumors in animals. In combination with potent chemotherapeutic drugs, they have led to shrinking of a tumor and in some cases to its complete disappearance. Clinical trials of antisense drugs are already under way. Will the antisense drugs one day create a hoopla comparable to the one that occurred in 1996 with the HIV protease inhibitors? The data on animals are encouraging enough to hope that this may happen.

The second strategy that is attracting a growing interest, is based on DNA's ability to form triplexes, which we discussed in Chapter 11 in connection with the H form. In the H form, a triplex is formed intramolecularly, because of DNA's peculiar folding. It is clear, however, that if we add to the homopurine–homopyrimidine duplex—the sequence $(CT)_n \cdot (AG)_n$, for example—an appropriate homopyrimidine oligonucleotide, in this case $(CT)_m$, then, with acid pH, a strong and specific complex will arise. In fact, the possibility of forming such intermolecular triplexes was demonstrated for artificial nucleotide acids (RNA and DNA) long before the discovery of the H form: A. Rich, G. Felsenfeld, and D. Davies first demonstrated back in 1957 the formation of a triplex by artificial RNA molecules. Triplex research received a strong impetus after the discovery of the H form, when it turned out that in certain conditions, sections of natural DNA could fold themselves into a triplex. Working in parallel in several laboratories, Peter Dervan at Cal Tech, Claude Helenè in Paris, and researchers in our laboratory in Moscow, demonstrated through different methods that pyrimidine oligonucleotides were capable of recognizing appropriate homopurine–homopyrimidine sequences and could form strong complexes with them, especially under acidic conditions.

The result was the idea of a triplex strategy for the purposeful modification of the genome. The assumption was that the oligonucleotides themselves, thanks to the formation of a triplex, could block the functioning of the genes or that they could be used to deliver to DNA sections the chemically active groups capable of modifying DNA (e.g., cutting DNA).

The triplex strategy looks especially attractive when applied to the treatment of cancer. Let us imagine that we have been able to choose an oligonucleotide that binds strongly only with the oncogene. This oligonucleotide, carrying a chemical group capable of cutting DNA, penetrates the cell's nucleus and binds with the appropriate DNA section. Thus, the oligonucleotide would constitute an ideal chemotherapeutic means capable of selectively destroying only oncogene-carrying cells. The road to the strategy's implementation, however, is strewn with many difficulties, not all of which can be regarded as resolved, even in principle. Some of these problems (and possible solutions) are as follows.

First, pyrimidine oligonucleotides form a strong complex only under acid pH, so it is generally difficult to hope that such a triplex will form inside the cell. It has proved possible to overcome this difficulty in a number of ways. The purine oligonucleotides are capable of forming triplexes with homopurine–homopyrimidine sections at neutral pH. These triplexes are stabilized by ions of different metals that the cell has available. In addition, chemists have synthesized several artificial analogs of nitrogenous bases (not found in nature) that form quite strong triplexes with DNA under physiological conditions.

Second, to get to DNA, the oligonucleotide must penetrate the cell's nucleus. This problem has turned out to be very difficult to resolve. There is hope, however, that, with a vast range of possibilities and using different chemical modifications of oligonucleotide, it will be overcome.

Third, the oligonucleotide must not be simply "gobbled up" by cellular nucleases en route to DNA. Oligonucleotide's chemical modifications are of great help with this problem, because they make the oligonucleotide "inedible" for nucleases.

Fourth, instead of binding with DNA, oligonucleotide may enter into an "illegal liaison" with many of the cell's RNA molecules, forming with them not a triplex but an ordinary duplex, albeit not for the whole of its length. This possibility has yet to be studied thoroughly. However, the successes of the antisense strategy, in which "illegal liaisons" have not materialized, give us grounds for hoping that this danger will not be too great for the triplex strategy.

The future will show whether it will be possible to clear all the hurdles on the road to a triplex strategy and whether the drugstores will eventually sell miraculous "genetic" medicines based on oligonucleotides. In the meantime, researchers are moving boldly. In 1991, a team from Copenhagen University (Peter Nielsen, Michael Egholm, Rolf Berg, and Ole Burchardt) tried a radical approach to the chemical engineering of DNA, to produce a molecule remotely reminiscent of DNA, but carrying, instead of the sugar–phosphate backbone, a totally different chemical group, reminiscent of a protein-molecule backbone. They called the molecule, schematically shown in Figure 37a, PNA (i.e., peptide nucleic acid).

The molecule has a completely different design from any other known to nature or chemists. The concept was to have distances between nitrogenous bases in PNA be the same as those in DNA. The authors hoped that such a molecule, being uncharged (unlike nucleic acid, which carries a single negative charge in each nucleotide—see Chapter 6), would form a very strong triplex with the DNA duplex. Therefore, they added PNA-T_{10} to DNA carrying an $A_{10} \cdot T_{10}$ insert. Imagine their amazement when instead of forming a triplex, PNA-T_{10} ousted the T_{10} strand from DNA and apparently formed a structure such as that shown in Figure 37c.

Further research revealed that a triplex was indeed formed, but not of the type originally expected (i.e., one PNA molecule with a DNA duplex). Instead, the structure formed consisted of two PNA molecules forming a triplex with a single DNA strand. The shape of this triplex is so handy that, to make it occur, the complementary DNA strands diverge and one of them forms a triplex with two PNA molecules (Figure 37c).

The triplex formed by two homopyrimidine PNA molecules and one homopurine DNA strand is exceptionally stable. Apparently, the unusual

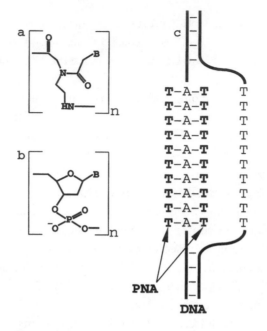

Figure 37. The structure of (a) peptide nucleic acid (PNA), in which the (b) sugar–phosphate backbone of DNA is replaced into a polyamide chain; only the base (B) remains the same. (c) Strand displacement of the T strand of linear plasmid DNA by oligo-PNA.Note that not one but two PNA molecules are bound to the A strand, forming the triplex.

stability of this triplex is due to two factors. First, there is no electrostatic repulsion within the three-stranded complex. Such repulsion significantly de-stabilizes the canonical DNA triplexes consisting of three strongly charged DNA strands. Secondly, an X-ray crystallographic structure of the $(PNA)_2$/ DNA triplex (solved by Steven Jordan and his co-workers at Glaxo) revealed additional hydrogen bonds between phosphate groups of the DNA backbone and nitrogens of the PNA backbone, belonging to the PNA molecule that plays the role of the third strand in the complex. Since duplex DNA is targeted by two PNA molecules, dimeric PNAs, or PNA clamps, were prepared, which consisted of two PNA molecules connected by a flexible linker. These PNA clamps bind to duplex DNA much more efficiently than monomeric PNAs.

In a collaborative study between my laboratory at Boston University and Peter Nielsen's laboratory at Copenhagen University, Vadim Demidov and Alexei Veselkov demonstrated that the exceptional stability of a complex of PNA with duplex DNA is accompanied by remarkable specificity of the interaction. This seemed strange, because if the complex is very strong and is stabilized, in a significant part, by non-sequence-specific backbone–backbone interactions, one must expect poor sequence specificity. This apparent contradiction has been resolved by our model of the two-stage binding of PNA to duplex DNA. During the first searching stage, one PNA molecule in the PNA clamp forms a highly unstable, transient duplex with the complementary strand of duplex DNA displacing the second DNA strand. In the second locking stage, a virtually irreversible triplex formation occurs. Thus, during the first stage, a high sequence-specificity of recognition is reached, whereas during the second stage, a remarkably stable complex is formed.

Therefore, we can speak of the third, a PNA strategy for gene-drug development. The practical utilization of PNA requires resolving at least part of the problems related to the triplex strategy using oligonucleotides. At least one problem does not exist for PNA: It is "inedible" for nucleases. Due to all of the above, the possibility of using PNA as both an antisense and antigene drug has, quite naturally, attracted a great deal of interest from the pharmaceutical industry.

AFTERWORD

SINCE THE DISCOVERY OF THE DNA DOUBLE HELIX IN 1953, DNA has been realized as the most important molecule of life. An enormous body of knowledge about DNA structure and functions has been accumulated. Efficient methods of manipulating DNA have been developed, among them methods of DNA cloning, sequencing, and amplification. These achievements have opened ways for bridging the molecular level (the level of DNA and its protein products) with the system level of the entire organism. As a result, we witness an unprecedented progress in understanding the molecular mechanisms of most basic processes in living organisms at norm and pathology, from mechanisms of cancer to brain functioning.

In parallel with these breakthroughs in basic research, we witness massive penetration of DNA technology into applied fields. Countless biotechnology companies exploit unheard-of opportunities to develop new diagnostics and treatments of virtually any disease. Among the most spectacular successes in this area are drugs that lower cholesterol for prevention of atherosclerosis, and protease inhibitors for preventing AIDS.

DNA technology has already revolutionized criminology. No serious crime is currently investigated without DNA analysis.

All these sweeping changes fascinate our imagination. And yet what we have witnessed so far will be seen in the near future as a very slow start. As the Human Genome Project approaches its triumphant completion, researchers determine the sequence and function of thousands of new human genes. More powerful and less expensive methods of diagnostics of alterations of DNA sequence are being developed. Today, very few hereditary diseases are diagnosed on the DNA level. Very soon, DNA diagnostics will become as routine a proceudre as the regular blood test. The treatment of diseases will rely more and more on the knowledge of DNA sequence and structure. Chemically modified DNA molecules will be used as drugs. The first such drugs, antisense oligonucleotides, are expected to be approved by the FDA in 1997, and will be used to treat patients. I am convinced that decisive breakthroughs in the battle with cancer are around the corner.

On the basic front, the most fascinating discoveries must be expected in the field of aging. In the coming years, we will learn a lot about the genetic factors that determine the "natural" life span in humans. In parallel with further progress in treatment of the major causes of "unnatural" death, atherosclerosis and cancer, an extension of humans' natural life span will eventually become a reality.

GLOSSARY

ADENINE: a chemical group that is part of DNA and RNA. One of the four bases of these nucleic acids. The abbreviated designation is A.

AIDS: acquired immunodeficiency syndrome. A very grave disease caused by an RNA-containing virus called HIV. The virus strikes at T lymphocytes; as a result, the patient loses the capacity for immune response.

ALLELE: one of the two genes responsible for a specific genetic characteristic. One allele comes from one parent; the other allele comes from the other parent.

ALLERGY: an excessive response of the immune system.

AMINO ACID: a chemical compound of the H2N - CHR - COOH structure, where R is any radical. An amino acid is the starting product for protein synthesis.

AMINO ACID RESIDUE: a chemical group of the - HN - CHR - CO - structure, which is a residue of the protein chain. It is what remains of an amino acid after it is built into the protein chain.

ANTIBIOTIC: an organic substance that suppresses multiplication of bacteria, but that is not poisonous to humans or animals. The first antibiotic was penicillin, isolated by Alexander Fleming from fungi in 1929. The discovery of antibiotics produced a revolutionary breakthrough in the treatment of many diseases that had previously resisted any medical treatment, such as pneumonia and tuberculosis. However, the broad and uncontrolled reliance on antibiotics caused bacteria to develop a resistance to them. As a result, traditional antibiotics are now much less effective than they were during the first decades of their application. Antibiotics are utterly irrelevant in combatting viral diseases.

ANTIBODY: a protein produced by the immune system in response to penetration of the organism by an alien substance—an antigen. Synonymous with *immunoglobulin*.

ANTIGEN: an alien substance that causes an immune reaction of the organism.

APOPTOSIS: programmed cell death. The phenomenon and its significance are discussed in Chapter 12.

ATHEROSCLEROSIS: a chronic ailment caused by a shrinking of blood-carrying vessels, mainly through formation on their internal walls of plaque that consists mostly of cholesterol.

ATP: the abbreviated name of adenosine triphosphoric acid. It is a universal accumulator of energy in the cell. Energy is stored in the triphosphate "tail" of the molecule. The "discharge" occurs as a result of separation of one phosphate group. The "charging" is carried out in mitochondria.

AZT: the abbreviation for $3' = $ azido $ = 2'$, $3' = $ dideoxythymidine a drug inhibiting reverse transcriptase. It is widely used against AIDS.

BACTERIA: single-cell microorganisms. The world of bacteria is extraordinarily varied and plays an immense role in ensuring the existence of other living things on earth. Many bacteria survive in the most primitive conditions, requiring for their multiplication only the simplest molecules containing chemical elements that are part of biological molecules. Thus, to meet their carbon needs, some bacteria need only petroleum; they get their nitrogen and oxygen from the air. Bacteria are everywhere: they cause the souring of milk or broth; they dwell inside us, helping us to digest food (*E. Coli*); and they also cause many infectious diseases.

BACTERIOPHAGE: a virus that kills a bacterium. It consists of nucleic acid (DNA or RNA), placed within a protein coat. Infection of bacteria occurs when the bacteriophage sticks to the bacterium's surface and injects its nucleic acid into bacterium. Soon after this, the bacterium's resources switch over to synthesizing a viral nucleic acid and proteins. Within about twenty minutes of the infection, the bacterial membrane bursts open and spills out about 100 "ready-made" virus particles that are perfect copies of the original bacteriophage.

BACTERIORHODOPSIN: a protein in some bacteria, which plays a key role in transforming light into ATP.

BASE (NUCLEIC OR NITROGENOUS):   a class of chemical compounds that includes adenine, guanine, thymine, cytosine, and uracil.

BROWNIAN MOTION: a chaotic motion of microparticles suspended in a liquid, as a consequence of thermal agitation of molecules.

CARCINOGEN: an agent that causes cancer.

CATENANE: two or more circular molecules topologically linked with each other.

CHAPERONS: specialized proteins that carry out, inside the cell, the folding of a polyamino acid chain (newly synthesized on ribosome) into a native protein molecule.

CHIMERICAL DNA: an artificial molecule put together through genetic engineering techniques from sections of different natural DNA. The terms *hybrid* and *recombinant* DNA have the same meaning.

CHOLESTEROL: a complex organic molecule of the steroid class. In moderate quantities, it is necessary for building the cell membrane and serves as the predecessor for a number of hormones (including sex hormones). Excess cholesterol in blood leads to atherosclerosis.

CHROMOSOME: a complex-structured set of DNA with proteins inside the cell nucleus. It stores genetic information.

CLONING: obtaining a large number of cells from the same cell. Now also used in relation to DNA molecules.

CODE (GENETIC): a "dictionary" for translating DNA and RNA texts into the protein (amino acid) language.

CODON: a term connected with the genetic code. Denotes trinucleotides corresponding to one amino acid residue. There are several meaningless (nonsense) codons that do not correspond to any amino acid. They play the role of stop signals during protein synthesis by messenger RNA on a ribosome, and are called terminating codons. Initiating codons serve as signals for starting protein synthesis.

COIL (POLYMER): a notion of polymer physics that denotes the form a polymer molecule assumes in space. Because of thermal motion, the form of a polymer coil constantly changes.

COLLAGEN: a protein of connecting tissue. A major example of a protein that, not being an enzyme, plays a structural role. Collagen is the principal component of bones and tendons. In everyday life, it is known as *gelatin*. It is used to make things such as jellies, glue, and gelatins.

COMPLEMENTARITY: the property of DNA's double helix, responsible for adenine always matched with thymine (and vice versa), and for guanine always matched with cytosine (and vice versa).

COVALENT BOND: strong chemical bonds that ensure molecules' integrity (e.g., the bond in molecules H_2, N_2, CO, etc.).

CRUCIFORM: a DNA structure that can be formed in palindrome sequences.

CYTOPLASM: the content of a cell without the nucleus.

CYTOSINE: a chemical group that is part of DNA and RNA. One of the four bases of these nucleic acids. The abbreviated designation is C.

DEGENERATION OF THE CODE: one of the properties of the genetic code, which is that several codons may correspond to the same amino acid.

DEOXYRIBONUCLEIC ACID: the full name of the DNA molecule (see DNA).

DIABETES: a disease characterized by the accumulation of sugar in the blood, due to the inability of the pancreas to produce the insulin hormone.

DIFFERENTIATION: specialization of cells in the process of the development of a multicellular organism. The reason for the formation of specialized cells—such as cells of skin, blood, and so on—is one of the crucial problems of science remaining to be solved.

DNA: deoxyribonucleic acid. A molecule that contains genetic information. It consists of two polynucleotide strands that a form double helix. Linear DNA has two ends.Closed circular (ccDNA) has no ends. Either of the polynucleotide chains in ccDNA is closed unto itself. A single-stranded DNA consists of one polynucleotide chain.

DNA POLYMERASE: an enzyme responsible for DNA synthesis on the DNA template. The process is called DNA replication.

ELECTROPHORESIS: the motion of molecules in an electric field.

ENDONUCLEASE: an enzyme that splits nucleic acid at a random place in the chain, not only from the end, as does exonuclease.

ENTROPY: a quantitative measure of the degree of disorder of a system.

ENZYME: a protein molecule responsible for a specific chemical reaction in a cell. Enzymes are catalysts, that is, without changing themselves in the course of a reaction, they strongly accelerate the progress of the reaction.

Enzymes also ensure a very high specificity and selectiveness of reactions occurring in the cell.

ENZYMOLOGY: the science of enzymes.

ESCHERICHIA COLI: a bacterium living in natural conditions in man's bowels. Abbreviated as *E. coli*. Frequently used in research by molecular biologists.

EUKARYOTES: organisms having a cell nucleus.

EXON: a DNA segment that stores information about a part of the amino acid sequence of protein.

EXONUCLEASE: an enzyme that splits nucleic acid from the ends, nucleotide after nucleotide.

GEL: a polymer network saturated with a solvent. Like a solid substance, a gel retains its form (e.g., gelatins and jellies). Electrophoresis in gels is widely used in decoding DNA sequences, in genetic engineering, and in investigating circular DNA.

GENE: the principal component of classical genetics, which at first was, for a long time taken to mean an indivisible particle of heredity. Then, during the 1950s and 1960s, the word gene became synonymous with a continuous DNA section, on which information, in the form of nucleotide sequences, is recorded about the amino acid sequence of one protein. Now, after the discoveries discussed in Chapters 6 and 7, the definition of a gene has ceased to be so narrow. The word is still used as the name for a DNA segment, but in some cases it denotes a continuous segment that corresponds to only a part of the protein chain, whereas in others it means a set of segments corresponding to a complete protein molecule. It may well be that the same DNA segment simultaneously belongs to two and even three genes.

GENETIC ENGINEERING: an applied branch of molecular biology that engages in purposeful modification of heredity by cutting and "stitching together" DNA molecules and subsequently building them into a living cell.

GENETICS: the science of heredity.

GENOME: the entirety of genetic information of the organism.

GENOTYPE: a term from classical genetics, denoting the totality of genes of the given organism. Now, the term *genome*, which has the same meaning, is used more frequently.

GUANINE: a chemical group that is part of DNA and RNA. One of the four bases of these nucleic acids. The abbreviated designation is G.

HEART ATTACK: necrosis of part of the heart muscle as a result of disrupted blood circulation. One of the principal causes of human death. Frequently caused by clotting in the heart's blood-carrying vessel because of advanced atherosclerosis.

HELIX: a screwlike structure.

HEMOGLOBIN: a protein that transmits oxygen in the blood. It gives blood its red color.

HEPATITIS: a grave virus-induced disease of the liver.

HISTONES: proteins that are part of chromosomes. They form the protein core of nucleosomes.

HIV: human immunodeficiency virus. The name of the virus causing AIDS.

HOMOZYGOTICITY: a term in classical genetics denoting that alleles are identical in their manifestation.

HORMONES: molecules of protein and other origin that regulate many processes in the organism. The lack or excess of a particular hormone causes many chronic ailments. Such hormones as insulin, growth hormone, and others are widely known.

HYBRID DNA: an artificial molecule put together through genetic engineering techniques from segments of different natural DNA. The terms *recombinant* and *chimerical* DNA have the same meaning.

IMMUNITY: resistance to a particular infectious disease by one who has suffered it in the past. Research into immunity has led to the discovery of the immune system, which removes intruding alien substances from the organism.

IMMUNOGLOBULIN: a protein produced by the immune system in response to penetration of the organism by an alien substance. Often called *antibody*.

INOCULATION: introduction of a vaccine into an organism to produce immunity against a disease.

INSULIN: a protein hormone produced by the pancreas, which regulates sugar content in the blood.

INTERFERON: a protein produced in the organism in response to a viral infection. Differs from immunoglobulin and has nothing to do with the immune system. Effective against the most diverse viruses, it is among the most promising antiviral preparations. Its production in large quantities has become possible through genetic engineering.

INTERLEUKIN-2: a protein that is the growth factor of T lymphocytes.

INTRON: a DNA section that divides exons.

ISOLEUCINE: one of the twenty canonical amino acids.

LDL (LOW-DENSITY LIPOPROTEINS): fat corpuscles that serve to transfer cholesterol molecules.

LIGASE: an enzyme that joins together ruptures in DNA.

LINK: topological state of two or more contours, which cannot be taken apart without at least one of them breaking.

LINKING NUMBER: a quantitative characteristic of the degree to which two contours are linked. It equals the number of times that one contour pierces the surface spread tight over the other contour. Denoted as *Lk*.

LYMPHOCYTES: blood cells responsible for immunity.

MALIGNANT: oncogenic. The process of emergence of a cancerous tumor is called *malignant degeneration* of tissue.

METASTASIS: a secondary pocket of malignant degeneration. An indication of an advanced stage of cancer.

METHIONINE: one of the twenty canonical amino acids.

METHYLASE: an enzyme responsible for methylation.

METHYLATION: addition of the methyl group CH_3.

MITOCHONDRION: a cigar-shaped body located in the cytoplasm. It is the cell's power plant, transforming food products into adenosine triphosphoric acid (ATP) energy.

MUTAGEN: an agent causing mutation.

MUTATION: an inheritable change in genetic material. Mutations may be spontaneous (i.e., produced by natural causes) or induced (i.e., triggered artificially, by radiation, chemicals, and so on). Mutation results in a change in DNA's nucleotide sequence.

NANOMETER (NM): a unit for measuring length (a billionth part of a meter, $1 \text{ nm} = 10^{-9} \text{ m}$).

NEWLY ACQUIRED CHARACTERISTIC: a characteristic not inherited, but appearing under external influences. Newly acquired characteristics are not inherited because they are not reflected in genes. For instance, no matter how exotic the color of the substance hair dye, it will not be reflected in the hair color of future progeny.

NUCLEASE: an enzyme that splits DNA or RNA.

NUCLEIC ACIDS: DNA and RNA.

NUCLEOSOME: the principal structural element of the chromosome. It consists of a protein (histone) core, on which is wound DNA with a length of 140 base pairs, thus making about two turns.

NUCLEOTIDE: a residue of DNA and RNA.

OLIGONUCLEOTIDE: a short, single-stranded piece of nucleic acid.

ONCOGENE: a gene that causes cancer.

ONCOGENIC VIRUS: a virus that causes cancer.

OVUM: a female sex cell.

PALINDROME: a phrase that reads the same from left to right or from right to left. In the written palindrome the spaces between words and punctuation marks are not taken into account. For example: "Madam, I'm Adam." In the context of DNA, such palindromes are called *mirror repeats*. A real DNA palindrome is a segment of double helix that has the same sequence when either strand is read in the same direction, dictated by the chemical structure of DNA strands. For example:

$$\text{---------}>$$
$$\text{ATGCGCAT}$$
$$\cdot\ \cdot\ \cdot\ \cdot\ \cdot\ \cdot\ \cdot$$
$$\text{TACGCGTA}$$
$$<\text{---------}$$

PCR (POLYMERASE CHAIN REACTION): a technique that makes it possible to amplify in a test tube a chosen piece of DNA as described at length in Chapter 10. A major tool of genetic engineering and biotechnology.

PENICILLIN: the first antibiotic, discovered by Alexander Fleming.

PENICILLINASE: an enzyme that splits penicillin, making it inactive. Production of the enzyme by certain bacteria protects them from penicillin's effect.

PHAGE: the abbreviated name for *bacteriophage*.

PHASE: one of the three states of matter (solid, liquid, or gaseous).

PHASE TRANSITION: transition of a substance from one phase state into another.

PHENOTYPE: a notion of classical genetics signifying the totality of external characteristics and properties of a living organism, which have evolved in the course of its development.

PHENYLALANINE: one of the twenty canonical amino acids.

PHOSPHATE: a chemical group that is part of a nucleotide.

PHOTODIMER (OF THYMINE): a special chemical compound, formed after one of the two thymines standing together along the same strand in DNA has absorbed a photon.

PLASMID: a circular DNA molecule multiplying together with bacteria and capable of passing from cell to cell.

POLY-: an affix denoting a polymer.

POLYMER: a chemical compound representing a chain of recurring groups, the simplest of which is polyethylene:

$$\ldots - CH_2 - CH_2 - CH_2 - \ldots$$

used in making packaging, handbags, and many other things. Homopolymers consist of perfectly identical residues. Biological polymers are heteropolymers, since in each of them the residues, albeit belonging to the same class (amino acids in protein and nucleotides in nucleic acids), differ in their structure. A protein consists of residues of twenty kinds, nucleic acid consists of residues of four kinds.

PRIMER: a single-stranded oligonucleotide (DNA or RNA) that binds, via complementary pairing, to DNA or RNA single-stranded molecules and serves for the priming of polymerases working on both DNA and RNA.

PROGENITOR: the hypothetical common forefather of all living things on Earth.

PROKARYOTES: single-cell organisms, with no cell nuclei.

PROMOTER: a DNA segment with which RNA polymerase binds to initiate messenger RNA synthesis.

PROTEASE: an enzyme that splits proteins.

PROTEIN: a major component of the living cell. It is a polyamino acid chain forming an involved spatial structure. Natural proteins are heteropolymers consisting of amino acid residues of twenty kinds. One protein differs from another by the sequence of amino acid residues.

PURINE: a class of chemical compounds that includes adenine and guanine.

PYRIMIDINE: a class of chemical compounds that includes thymine, uracil, and cytosine.

RECEPTORS: protein molecules built into the cell membrane and responsible for the cell's reception of external signals. These signals are other protein molecules floating in intercellular medium. Examples include receptors of growth factors, receptors of antigens of T lymphocytes, and receptors of low-density lipoproteins (LDL) in liver cells.

RECESSIVENESS: in classical genetics, the notion that a recessive gene manifests itself only in a homozygotic state.

RECOMBINANT DNA: an artificial molecule put together through genetic engineering techniques from segments of different natural DNA. The terms *hybrid* and *chimeric* DNA have the same meaning.

REPAIR: the curing of damaged places in DNA, which is discussed at length in Chapter 3.

REPLICATION: doubling of genetic material.

REPRESSOR: a protein binding very strongly with an appropriate DNA segment between the promoter and the gene itself. By associating with DNA, repressor prevents the progress of RNA polymerase from promoter to gene and thus blocks messenger RNA synthesis. Serves to regulate transcription.

REPTATION: a snakelike movement of a polymer molecule through the polymer network. The notion is crucial for theoretical understanding of DNA separation in a gel.

RESIDUE: in polymers, a recurring unit in a polymer chain, for instance, the CH_2 group in polyethylene, the amino acid residue in protein, and the nucleotide in DNA.

RESTRICTION ENDONUCLEASE: an enzyme cutting the double helix in places with a definite nucleotide sequence. It is the chief tool of genetic engineering. Often called *restriction enzyme*.

RESTRICTION FRAGMENT: a piece of DNA that is cut out of the molecule with the help of restriction endonucleases.

REVERSE TRANSCRIPTASE: an enzyme responsible for DNA synthesis on the RNA template. The process is called *reverse transcription*.

RHODOPSIN: a protein molecule that plays a key role in transforming light into the visual signal in the eye.

RIBONUCLEIC ACID: the full name of the RNA molecule.

GLOSSARY

RIBOSOME: an involved complex of RNA and proteins, which is responsible for the translation process in the cell.

RIBOZYME: an RNA molecule that works as a catalyst.

RNA: ribonucleic acid. A biological polymer akin to DNA in its chemical structure. It is capable of forming a double helix, but in nature, as a rule, it exists as a single strand. In some viruses, it is the carrier of genetic information, that is, it supersedes DNA. RNA has no genetic role in the cell, but plays an important role in transmitting information from DNA to protein. Three RNA types are distinguished by the functions they perform: messenger (mRNA), ribosomal (rRNA), and transfer (tRNA).

RNA POLYMERASE: an enzyme synthesizing messenger RNA on the DNA template, which carries out the transcription process.

SEGMENT (KUHN OR STATISTICAL): in polymer physics, an element of an idealized polymer chain consisting of rectilinear segments linked by loose hinges.

SELECTIVE CONDITIONS: conditions in which only those bacteria that possess some special properties can multiply. For example, in a medium to which an antibiotic has been added, only those bacteria that carry the gene of resistance toward the antibiotic in question can multiply.

SPECIES: one of the principal notions of descriptive biology, which systematizes living things. The basic principle underlying the division into species is that of being able to produce fertile progeny by interbreeding. For instance, donkeys and horses belong to different species, because the product of their crossing (the mule) is incapable of breeding (i.e., sterile).

SPLICING: the process of messenger RNA maturing in eukaryotes, which results in the discharge of introns, while exons combine into one RNA chain.

STRAIN: a set of bacterial cells obtained from a single cell. The term *clone* is also used in the same sense.

SUGAR: a chemical group that is part of a nucleotide. It belongs to the same class of compounds as sugar used for food.

SUPERCOILING: a characteristic of a circular closed DNA. Supercoiling arises when the linking number Lk in DNA differs from the N/γ_o value, where N is the number of base pairs in DNA and γ_o is the number of base pairs per turn of the double helix in linear DNA in the same conditions. Synonymous with superhelicity.

SUPERHELICITY: synonymous with supercoiling.

SYMBIOSIS: a mutually advantageous coexistence of two or more species. Symbiosis is of enormous significance for all living things on Earth. It is because of symbiosis of nitrogen-fixing bacteria and tuberous plants, that nitrogen from the air gets into plants, where it is absolutely indispensable as an element included in proteins and nucleic acids. By themselves, plants are incapable of assimilating nitrogen from the air.

TEMPLATE: a polymer molecule whose sequence is used for charting a sequence for another polymer molecule. DNA serves as a template for DNA synthesis during replication and for RNA during transcription. RNA serves as a template for protein synthesis during translation and DNA during reverse transcription.

THYMINE: a chemical group that is part of the nucleic acid, DNA. One of the four DNA bases. The abbreviated designation is T.

TOPOISOMERASES: a class of enzymes that change the topology of a circular closed DNA.

TOPOISOMERS: DNA molecules, identical in chemical terms, but different in topology (by the type of knot or by the linking number).

TOPOLOGY: a branch of mathematics studying general properties of curves and surfaces that do not change under any type of deformations without cutting and gluing.

TRANSCRIPTION: RNA synthesis on a DNA template. *Reverse transcription* is DNA synthesis on an RNA template.

TRANSFORMATION: the change in the heredity of a cell penetrated by alien genetic material.

TRANSLATION: protein synthesis by a messenger RNA template on ribosome.

TRYPTOPHAN: one of the twenty canonical amino acids.

TWINS (IDENTICAL OR MONOZYGOTIC): two brothers or two sisters who have grown from one zygote. Twins result when, before the embryo begins to develop, the fertilized ovum cell divides into two zygotes, each of which gives rise to a separate embryo. Identical twins have precisely the same set of genes; thus, they always belong to one sex and look alike. A comprehensive study of identical twins separated at an early age yields a wealth of information about the role played by genes and the role played by external conditions.

ULTRAVIOLET (UV) RAYS: irradiation of an electromagnetic nature, not seen with the naked eye, with a wavelength under 400 nm.

URACIL: a chemical group that is part of the nucleic acid RNA. One of the four RNA bases. The abbreviated designation is U.

VACCINE: a preparation containing bacteria or viruses that have been rendered harmless (i.e., killed), used for inoculation of people against infectious diseases.

VALINE: one of the twenty canonical amino acids.

VECTOR: a genetic engineering term. This is the name of a carrier DNA molecule (plasmid, virus, etc.), within which the desired gene is cloned.

VIRUS: a cellular parasite, one of the simplest objects of living nature. Outside the cell, a virus is a molecular complex consisting of nucleic acid (DNA, sometimes RNA) and several proteins forming the virus coat. When a cell is penetrated by a virus (or its nucleic acid), the cell switches its resources over to synthesizing the virus's nucleic acid and proteins. When the cell's resources are exhausted, its membrane bursts, spilling out "ready-made" virus particles. Animal viruses have a much simpler structure than bacteria viruses (bacteriophages). Viruses cause many infectious diseases, such as influenza, smallpox, poliomyelitis, hepatitis, and AIDS. In some cases, once inside the cell, the virus does not undo it but builds its DNA into that of the cell, after which the viral DNA begins to multiply together with the cell DNA. The behavior of the cell itself may, however, undergo a sharp change.

WRITHING: a notion of the theory of ribbons. The writhing value depends only on the form that the ribbon's axis takes in space, not on how the ribbon is twisted around that axis. It is denoted as Wr.

X RAY: short-wave electromagnetic radiation with a wavelength on the order of 10^{-10} m.

X-RAY CRYSTALLOGRAPHY: a method of determining the internal structure of crystalline substances by special processing of the X-ray patterns resulting from them. This is the most direct and powerful method for determining the structure of a substance. Our knowledge about the structure of molecules of any complexity—including the principal biological molecules, proteins, and nucleic acids—was obtained through the use of this method.

X-RAY PATTERN: an image on a photo plate, resulting from its exposure to X rays scattered by a crystal.

ZYGOTE: a fertilized egg cell; the single cell from which the whole organism develops.

INDEX

Brenner, Sidney, 21
Bronstein, Matvei, 16
Brown, Michael, 181
Brownian motion, 41, 192
Burchardt, Ole, 185

Cancer
    DNA and, 169–176
    mechanism of, 168–170
    tumor necrosis factor and, 177–180
Carcinogenicity, 170, 192
    testing for, 172–174
Catenane, 115, 118, 120f, 192
Cavendish Laboratory, 7, 9
Cech, Thomas, 77
Cell, repair of, 32
Cell suicidal syndrome, 177
Cell-free system, 22, 25
Central dogma, 46
Cetus Corporation, 133
Chain reactions, 131
Chaperon proteins, 161–162, 192
Chargaff, Erwin, 13
Chemical robot, 160
Chimerical, defined, 51, 192
Cholesterol, 180, 181, 192–193
Chromosome, defined, 193
Cloning, 28
    DNA, defined, 56, 193
Closed circular DNA (ccDNA), 85–87, 88, 88f
    topology of, 96, 97–100
Code, genetic. See Genetic code
Codons, 21
    deciphering, 22–23
    defined, 193
    initiation, 23
    stop, 24f, 193
    termination, 23, 24f
Cohesive ends, 117
Coil, polymeric, 42, 43, 193
Collagen, defined, 193
Complementarity rule, 13, 14f, 193
Covalent bonds
    defined, 193
    in polymer chains, 14
Cozzarelli, N., 121

Crick, Francis, 6
    and DNA structure, 9–15, 18, 141, 146
    and genetic code, 21
Crossbreeding, 52
Crothers, D., 36
Crowell, R., 113
Cruciform DNA, 100, 102–104, 103f
    defined, 193
    formation of, 105, 107f
    role of, 107
Cytidine, 13
    structure of, 20f
Cytochrome oxidase, 68
Cytoplasm, 194
Cytosine, 194

Davies, D., 184
de Broglie, Louis, 6
de Gennes, P.-J., 61
Death, genetic programming of, 177
Degeneration of the code, 194
Delbrück, Max, 1, 2–5, 6, 14, 18, 176
Demidov, Vadim, 187
Deoxyribonucleic acid. See DNA
Dervan, P., 184
Deuterium, 30
Diabetes, 194
Dickerson, Richard, 148
Dideoxy-chain termination method, 64
Dideoxythymidine, 128
Differential geometry, 95–96
Differentiation, 194
DNA, 10, 194
    B form of, 142, 146, 148, 149f, 150, 151, 152
    base pairs of, 141, 142f
    bases of, 28
    breakage of, 47–48
    and cancer, 169–176
    cardiac influences of, 180–181
    cruciform, 100, 102–107, 103f, 107f
    decoding, 62–65
    early research on, 11
    H form of, 152–157, 156f
    heat absorption of, 36–38, 37f
    importance of, 188
    knotted, 113–121

Index